Solid-State Microwave Devices

Artech House

Solid-State Microwave Devices

Thomas S. Laverghetta

Artech House   Boston • London

Library of Congress Cataloging-in-Publication Data

Laverghetta, Thomas S.
  Solid-state microwave devices.

  Bibliography: p.
  Includes index.
  1. Microwave devices.   2. Solid state electronics.
I. Title.
TK7876.L385 1987   621.381'3   87-19319
ISBN 0-89006-216-1

Copyright © 1987

ARTECH HOUSE, INC.
685 Canton Street
Norwood, MA 02062

International Standard Book Number: 0-89006-216-1
Library of Congress Catalog Card Number: 87-19319

10 9 8 7 6 5 4 3 2 1

To My Daughter Renee, My Little Cassy, Whom I Love Very Much and of Whom I Am Very Proud

Contents

Preface

The area of microwave solid-state devices is one about which many people like to talk, but one few people truly understand. Such questions as what are the characteristics of the materials used in solid-state devices; how does a PN junction or a Schottky junction actually operate; or why is a high electron mobility transistor different from GaAs FETs or bipolars should be areas of knowledge for the true solid-state designer. This book was written to give the designer that information in down-to-earth, understandable language.

Chapter 1 provides the reader with certain terminology and compares the solid-state devices of the 1970s with those available today. Chapter 2 presents detailed information on the materials used to fabricate microwave diodes and transistors. The materials covered are germanium, silicon, gallium arsenide, and indium phosphide.

Chapter 3 covers junctions used in solid-state devices. Detailed discussions of the PN junction and the Schottky junction are presented. Conditions of no bias, forward, and reverse bias are covered. Also provided are definitions of junction capacitance and transit time.

Chapter 4 is devoted to the microwave diode. Six types are covered including Schottky, PIN, varactor, Gunn, IMPATT, and TRAPATT. Chapter 5 completes the device presentation by covering bipolar transistors, GaAs FETs, and HEMTs.

Chapter 6 ties all the devices together by presenting the components fabricated from them. Components such as amplifiers, oscillators, attenuators, switches, detectors, mixers, and phase shifters are covered in design examples.

With the information presented here, the microwave designer will have the materials necessary for very admirable designs of the solid-state components needed in any system or application.

Thomas S. Laverghetta
Auburn, Indiana
May 1987

Acknowledgements

Many people are to be thanked for their contributions to this book. People who have provided a magazine, or a book, or even an informational word all have my heartfelt thanks and gratitude.

As always, there are those who have provided the extra effort that makes such an undertaking so much easier. To Beth Krueger who did the typing at virtually the last minute and did an excellent job; to Joanne Zelle who once again produced drawings I was proud to submit; and to Ron Emery who gave me encouragement, support, and my own private room to write in — a large THANK YOU.

Finally, to my wife, Pat, and to my family whose love and understanding helped tremendously on the days when nothing I had written looked right or seemed to want to fall into place — Thank you for being there.

Acknowledgments

Chapter 1
Introduction

The term *solid-state* in microwaves for a long time meant that you were limited as to how high in frequency you could operate. It was thought, and accepted, that when active requirements were presented in a microwave area, tubes were the only practical devices to use. For many years this actually was a reality. Even after the transistor had been shown to operate satisfactorily at some of the low microwave frequencies, many still clung to their old reliable tubes. Fortunately the transistor and the diode have found their way into many microwave areas and, in many cases, are the heart and soul of numerous microwave systems.

When we speak of solid-state devices we are speaking of the two components mentioned above, the transistor and the diode. To grasp how important they are to microwaves and to the entire microwave industry, we should look at each and see how each has progressed in a relatively short period of time. The time span to be considered begins in 1973. This was about the time when gallium arsenide field effect transistors (GaAs FETs) and diodes of GaAs construction were beginning to make an impact on the industry. The devices used in microwaves at that time will be compared to those available at the time this text is published.

To understand how the transistor has progressed over the years in microwaves, consider three areas: Low-noise, linear, and power devices. Figure 1.1 is a plot of noise figure *versus* frequency for the transistors available in 1973. These are galium arsenide field effect transistors (GaAs FETs) and this was about the best performance you could expect. The numbers on the ends of each plot are the associated gains for that particular noise figure.

Now consider Figure 1.2. This is a plot of the same parameters (noise figure *versus* frequency) for GaAs FETs in 1986. The area to note is the frequency limit that has been extended in Figure 1.2 to 40 GHz. The frequency range in Figure 1.1 stopped at 12 GHz. It is also interesting to note that the noise figure at 12 GHz in Figure 1.1 (3.5 dB) is nearly as good as the noise figure at 40 GHz in Figure 1.2 (3.6 dB). Also, the gain at 40 GHz in 1986 is better than the gain at 12 GHz in 1973 (6.2 dB as compared to

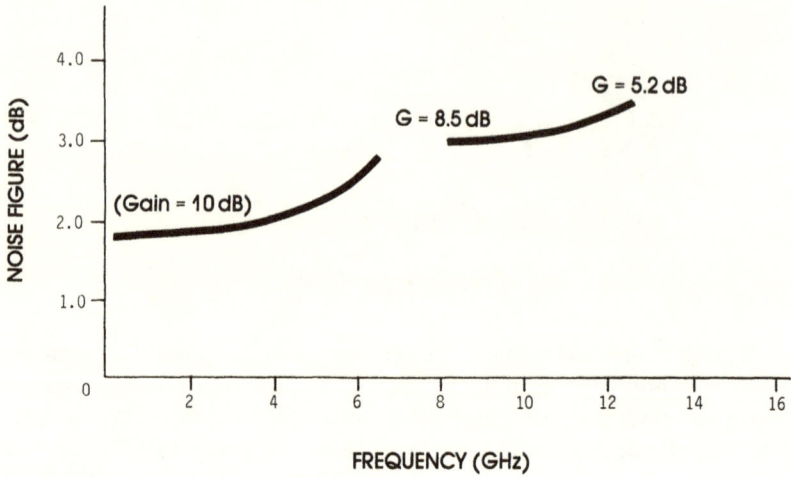

Figure 1.1 Low-Noise GaAs FETs

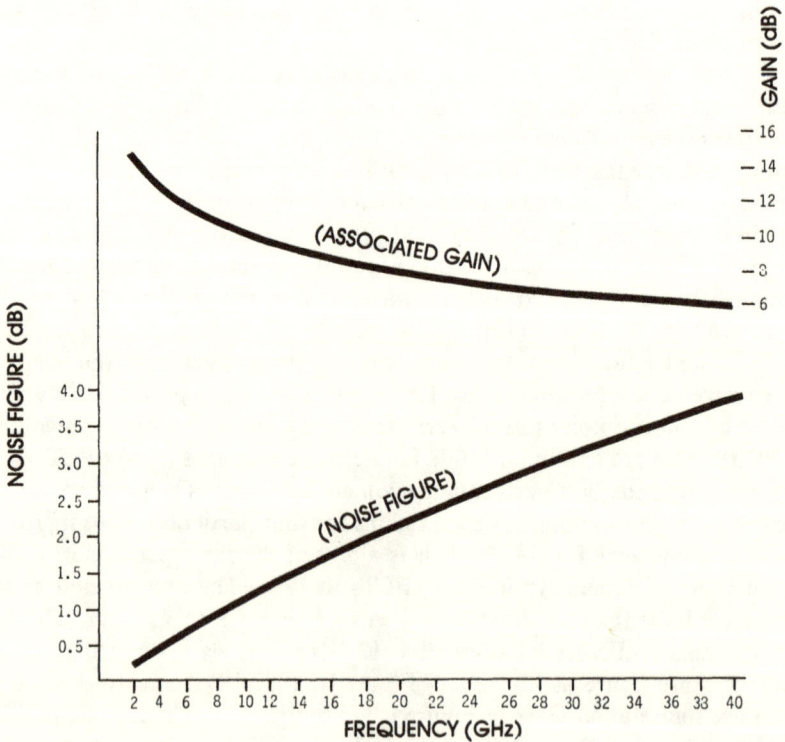

Figure 1.2 Low-Noise GaAs FETs 1986

5.2 dB). These two figures dramatically show the improvement in transistors for low-noise microwave applications.

The linear power devices (the class-A operation transistors) have not undergone such a dramatic change over the years as have the low-noise devices. Basically, the frequency range has remained the same with output power held constant. This can be seen by comparing Figures1.3 and 1.4. (It should be noted that one reason the frequency has not changed is that these are bipolar transistors in both cases. To achieve higher frequencies, GaAsFETs are needed, as will be explained in Chapter 5.) Figure 1.3 shows linear power devices operating up to 4 GHz in 1973 with output powers of 4 W (12 dB gain) at 1 GHz to approximately 1/8 W at 4 GHz with dB gain. The phenomenon to note in this figure is the rapid decrease in power output for each device as the frequency increases. These were truly narrow-band devices. Compare these devices to the ones in Figure 1.4 which are 1986 linear power devices. As previously stated, the frequency still goes to 4 GHz, but the power output is now much more consistent over a wider range of frequencies. The 4 W at 1 GHz now extends out to about 2.5 GHz and the 1/8 W at 4 GHz is now at 600 MW for the 1986 devices.

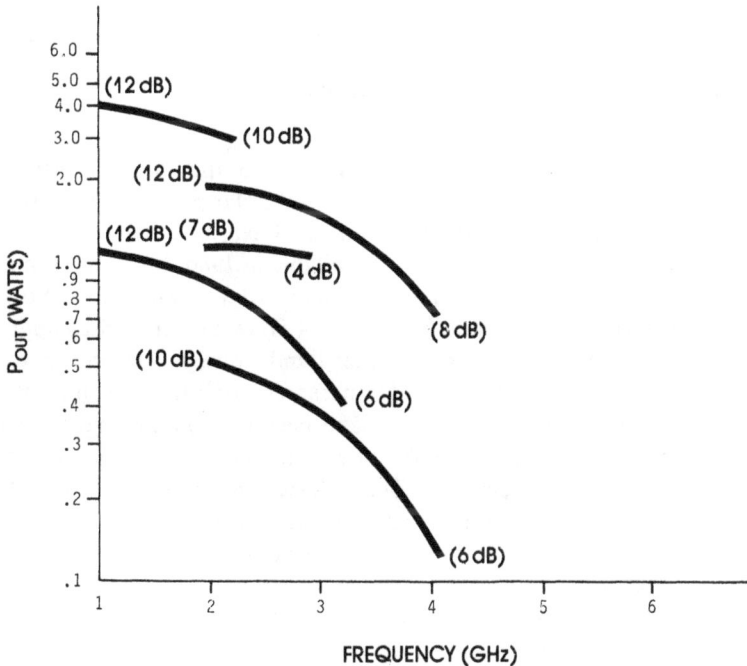

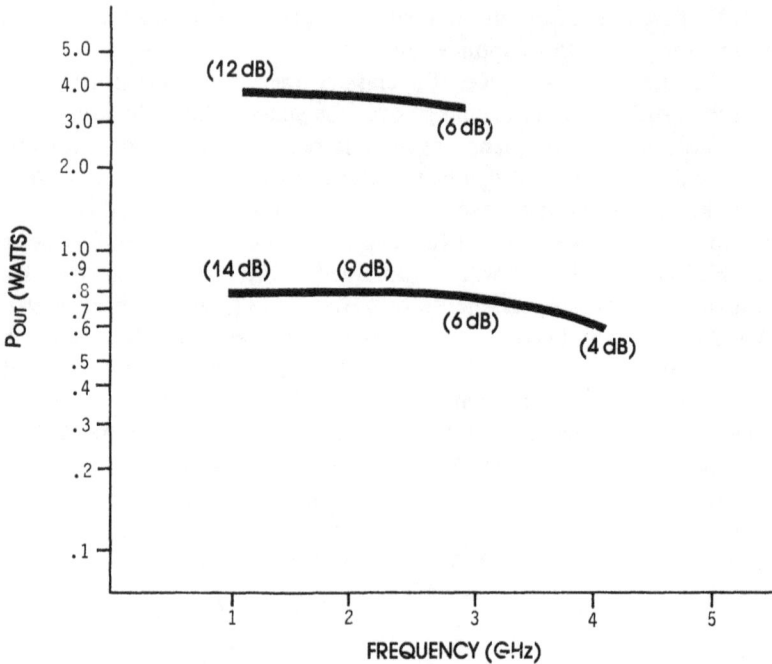

Figure 1.4 Linear Power Bipolar Transistors

Advances in power transistors between 1973 and 1986 are primarily in the area of Field Effect Transistors (FETs). Bipolar power transistors had their highest output powers between 1.0 and 2.0 GHz, as shown in Figure 1.5. Figure 1.6 shows the power transistors available in 1986. It can be seen that the lower frequencies have more capability in regard to power output. Powers of 40 W per device are available from 1.0–1.4 GHz, as opposed to approximately 30 W in 1973. FETs, on the other hand, were just beginning to be mentioned in 1973, and most of the references were for low-noise application. In Figure 1.6 you can see that they have become very important in the extension of the frequency range for power applications (as well as being very useful for low-noise circuits). These devices have extended operation so that 14–15 GHz power systems are realizable. It can be seen, therefore, that the transistor in microwaves has improved greatly in some areas and has simply made other areas (linear devices) more efficient and broadband.

The diode in microwaves has advanced similarly over the years. Using 1973 as a reference once again, it can be noted that the diode was used primarily for mixer and detector applications with the Gunn and IMPATT devices considered to be on the exotic side of the spectrum and specialized

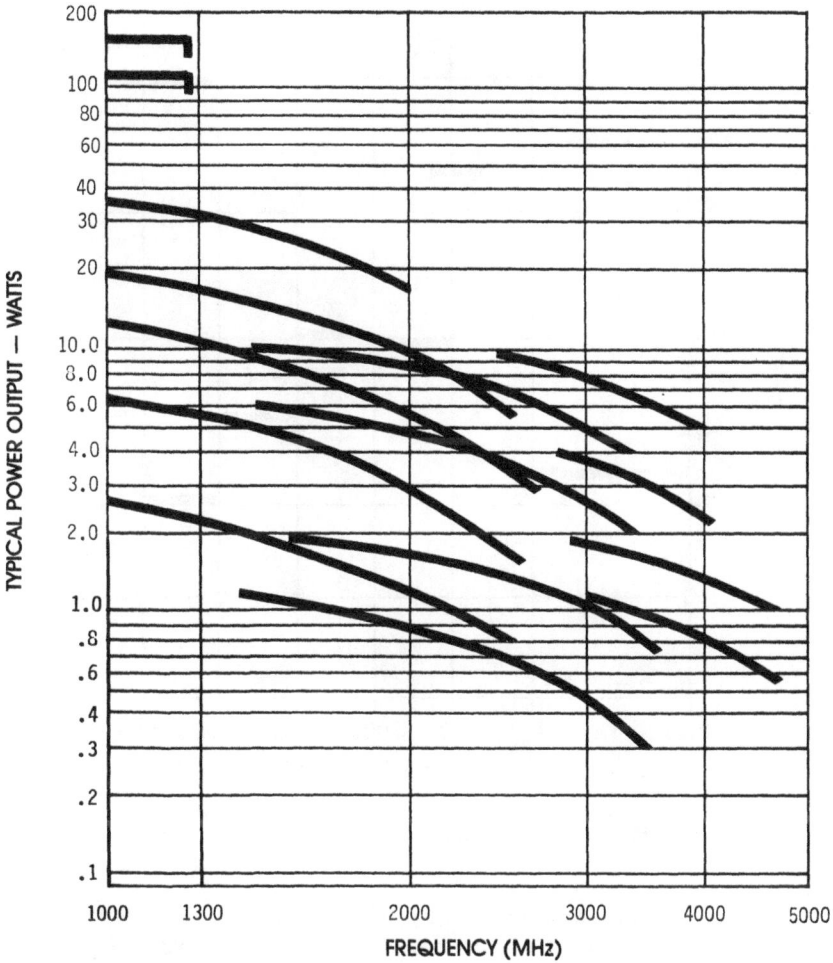

Figure 1.5 Power Transistors

devices. It had only been six years since Cayuga Associates had put their
LSA (limited space-charge accumulation) diode on the market which would
yield 100 W pulses at X Band (8–12 GHz). Typical operation of the so-called
transferred electron devices, were in the order of 200 W peak power at 9 GHz
with the majority of the devices capable of 80–100 W peak power in the
2–6 GHz range and 50 mW CW from 8–20 GHz. Avalanche devices, such
as IMPATT diodes, were capable of producing in the order of 10–20 W (both
pulse and CW) in the range of 8–15 GHz.

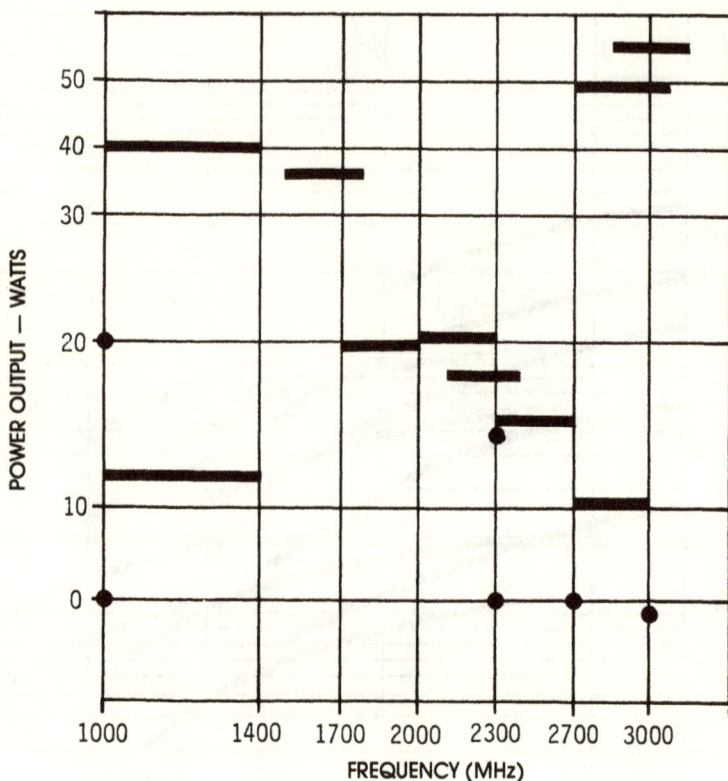

Figure 1.6 Power Transistors 1986

Present day Gunn diodes will produce output power comparable to those of 1973 with the addition of devices which will have CW power of 150 mW at 50 GHz and up to 20 mW at 94 GHz. The Gunn diode, as have most of the microwave diodes, has concentrated on extending the frequency of operation, rather than on increasing output power.

One application of microwave diodes which was very prominent in the early 1970s was as a tunnel diode amplifier. Satellite applications were increasing, and the need for devices which required very little power was also increasing. The tunnel diode amplifier (TDA) was just the unit for the job. It found many applications where low power drain was needed. Unfortunately, the tunnel diode has limited applications in the present-day scheme of things, since the GaAs FET has taken over in many cases. This is not because of low power consumption but because of size. The TDA required a circulator to operate in conjunction with the negative resistance of the tunnel diode. These can add considerable volume and weight to a circuit, and, in some cases, this can be prohibitive.

Thus it can be seen how solid-state devices have found wide application in the microwave field over the years. With the trend toward smaller and smaller circuitry to do more and more operations, semiconductors (transistor and diode) find many more uses than was ever dreamed possible.

So the term *solid-state* (defined as *pertaining to circuits and components using semiconductors*) takes on new meaning as we apply it to microwave and microwave applications. These tiny devices accomplish so many tasks they make the microwave industry remain a high technology profession. They can amplify, oscillate, mix, detect, attenuate, switch, modulate, phase-shift, and multiply microwave frequencies to such an extent that the founding fathers of microwave concepts would marvel at what they do. The microwave designer is in no way limited as to how high in frequency it is possible to operate. The choice now is which device is best for your application, *not*, is there one that will operate in your frequency range. *Microwaves* and *solid-state* are two terms that are definitely terms that belong together.

Chapter 2
Solid-State Materials

To appreciate fully and understand how, and why, microwave solid-state devices function as they do, it is necessary to investigate the materials used to fabricate them. An understanding of why germanium was used for early devices; why silicon was, and still is, the primary material for microwave devices; what makes gallium arsenide acceptable for even higher frequency applications; and what the compound indium phosphide is for the next generation of material for microwave transistors and diodes. All of these statements are part of the history of microwave solid-state devices and all of them involve a specific material. By knowing and recognizing each of these materials, an individual can make a much more intelligent choice when fulfilling a particular application.

The materials to be covered in this chapter have a variety of names. They are germanium, silicon, gallium arsenide, and indium phosphide. Each has distinct qualities which make it desirable for certain applications. Each has advantages and disadvantages. The one property they all have in common, however, is that they all are semiconductors. If you look in a dictionary you will see that a semiconductor is defined as:

> Material whose electrical resistivity is lower than that of insulators and higher than that of conductors.

What this definition is saying is that a semiconductor is not a conductor, not an insulator, but in between the two. To realize where semiconductors fall between conductors and insulators, consider that the typical conductor has a resistivity of 10^{-6} Ω cm; a typical insulator has a resistivity of 10^6 Ω cm; and a typical semiconductor has a resistivity of 10^1 Ω cm. So it can be seen that the semiconductor is actually in between conductors and insulators. That is, a true *semi*-conductor.

Just what is it that makes these materials semiconductors, and how do they lend themselves to microwave applications? To help answer these questions refer to Figure 2.1. This is a periodic table of the elements. Notice how the materials are all grouped together, being a part of either group 3, 4, or 5. The initial materials mentioned, germanium and silicon, are by

PERIODIC TABLE OF THE ELEMENTS

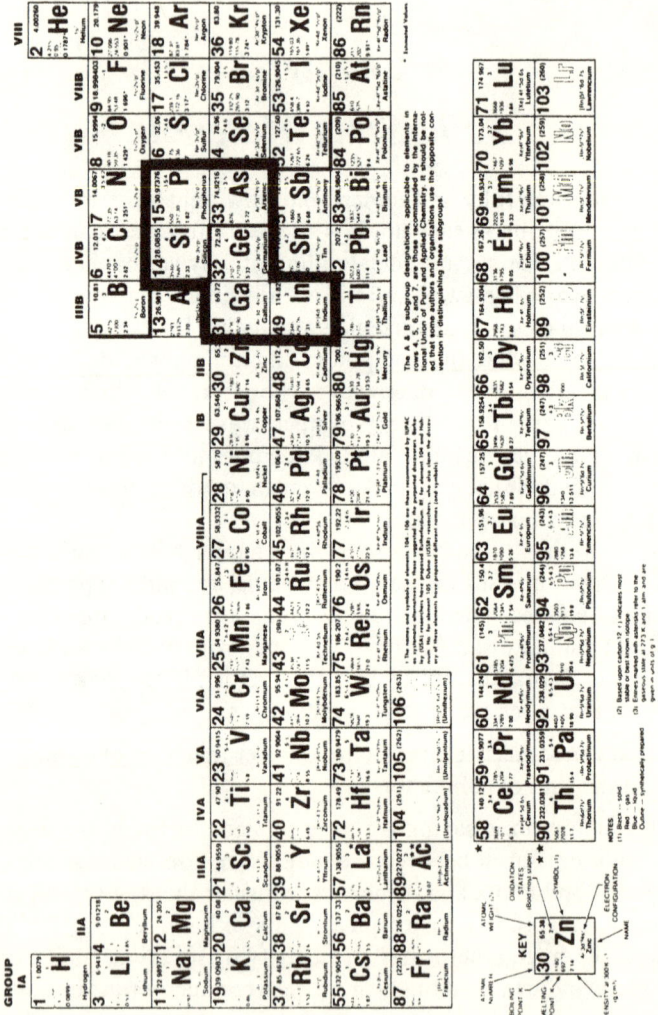

Figure 2.1 Periodic Table of the Elements

themselves in group 4. The compounds referred to initially, gallium arsenide and indium phosphide, consist of one element from group 3 (gallium and indium) and one from group 5 (arsenic and phosphorus).

The base materials, those in column four, are the materials which usually have impurities added to them. This is a process called *doping*. If, for example, arsenic atoms are added to germanium, there would still be covalent bonding between atoms, but there would be an excess electron as a result of the combination. This is shown in Figure 2.2. This is because arsenic has five electrons in its outermost orbit as opposed to the four which are present in germanium. This excess electron is termed a free electron. This condition decreases the resistivity of the overall combination and is termed a *donor* composition and is said to form *n*-type germanium. The *n* in this term is a designation for *negative*, which is the charge of the electron which is the primary current carrier in this case.

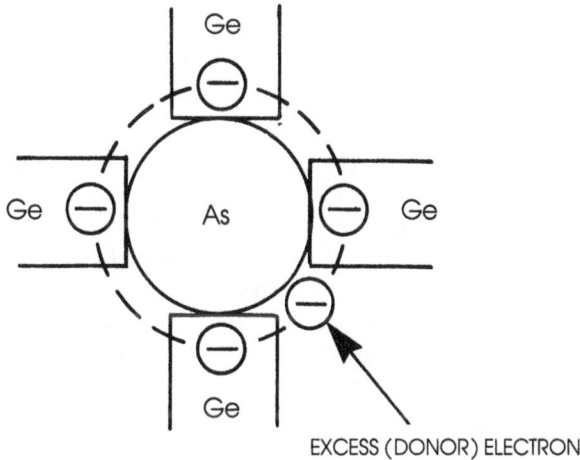

Figure 2.2 Donor Atoms

Similarly, if an atom from the third column was added to germanium, for example, there would be too few electrons available to make a covalent bond (Figure 2.3). In this case there would be a void, or *hole*, present. This construction constitutes a condition which is termed an *acceptor* atom. This means that the element formed (germanium and boron combination in this case) will accept an electron, if one is available. This condition is also termed a *p*-type material with the *p* representing a positive charge, which is the charge of the vacated *hole*. Once again, the resistivity of the germanium would be

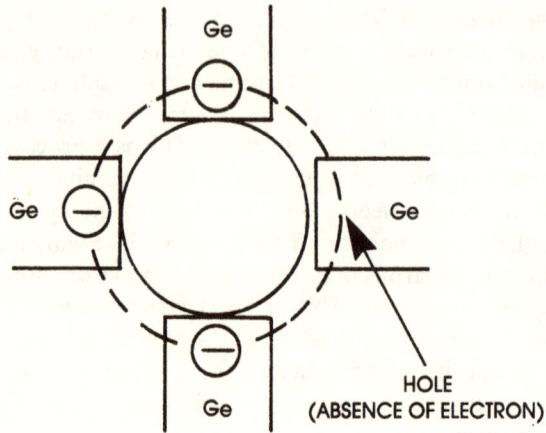

Figure 2.3 Acceptor Atoms

decreased, since the hole is an electrical carrier which can be considered comparable to a free electron when considering current flow.

Figure 2.4 shows this same relationship as gallium is combined with arsenic and indium with phosphorus. In each case an element from group three (gallium and indium) is combined with one from group five (arsenic and phosphorus). It can be seen that there is an excess of two electrons in this case, which makes these materials excellent majority carrier materials; that is, materials which have electrons as their source of current flow. There is no *hole movement* in such structures. These are called *unipolar* devices and will be covered extensively in later chapters.

The discussion above has shown the properties of the materials presented with regards to being semiconductors. To understand how each lends itself to microwave applications, it is necessary to look at each one individually. This is basically due to the fact that each material has unique properties which can be used to the highest degree for microwave applications.

2.1   GERMANIUM (Ge)

The material we begin with for our discussion is the material which actually started the whole science of semiconductors — *germanium*. This was the material used in 1948 by Brattain and Bardeen at Bell Telephone Laboratories, when they discovered the *point contact transistor*. This consisted of a small piece of germanium upon which metal contacts were connected. Experimentation found that when an input current was inserted

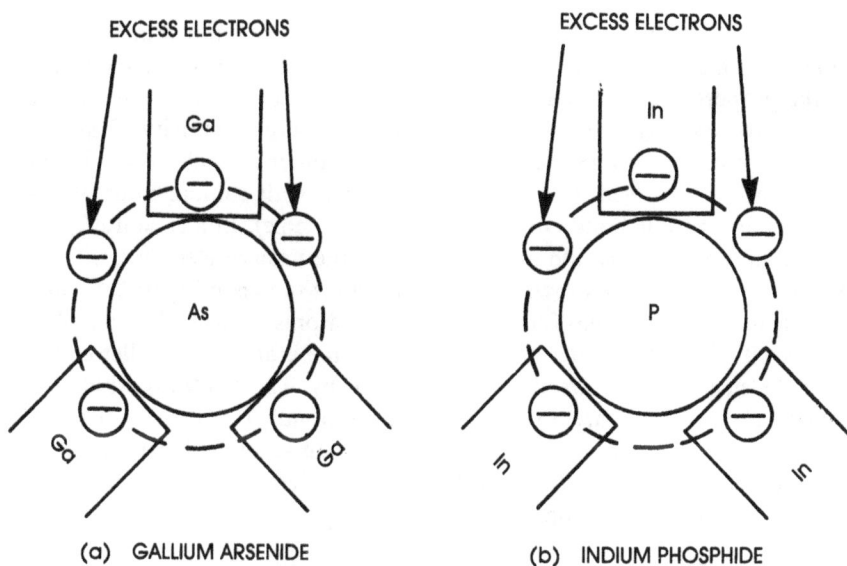

EXCESS ELECTRONS

Ga

As

Ga   Ga

(a) GALLIUM ARSENIDE

EXCESS ELECTRONS

In

P

In   In

(b) INDIUM PHOSPHIDE

Figure 2.4 Gallium Arsenide and Indium Phosphide

in the emitter contact, a larger current resulted at the collector contact (junction between one metal contact and the semiconductor). Although this first discovery had its drawback, the industry was on the threshold of the solid-state revolution. There in the middle of this revolution was germanium.

Although the year 1948 was referred to as the recognized beginning of the semiconductor movement, there was a great deal of research being done prior to that date. During the time period of 1942 to 1945, Karl Lark-Horovitz was involved in extensive germanium research at Purdue University. His early research concentrated on the resistivity and thermoelectric behavior of germanium semiconductors as well as on RF testing methods. By the end of World War II, the Purdue team of researchers had shown that the electrical properties of germanium could be predicted from its impurity content; that they could predict both resistivity and thermoelectric power of the crystal over a range of temperatures; that they had determined the mobility ratio for holes and electrons within the crystal; and that they had performed measurements which had determined that the dielectric constant of germanium was between 16 and 17. This was a tremendous amount of information to have available at that time on a material which until then had been considered to be one which could only be used for crystal rectifiers. Lark-Horovitz showed great foresight in choosing germanium for the first rectifier, a wise choice which produced an excellent semiconductor material.

As previously stated, germanium is in the fourth column of the periodic table. This means that there are four valence electrons in its outer shell. One of the properties a material must possess in order to be acceptable for use in microwaves is that it must have good electrical conductivity. This is a property possessed by germanium with its four outer shell electrons. It was noted earlier how well it conducts when joined with column three atoms (boron, for example) and column five atoms (arsenic). In a crystal of pure germanium each of the atoms shares one of the valence electrons with its neighbors. A simplified drawing of this concept is shown in Figure 2.5. Note that each of the atoms has four equally spaced atoms associated with it. The circles with Ge in the center are the nucleus of the atoms, the smaller circles with the negative sign are the valence electrons. The electrons which are touching one another are involved in covalent bonds. It is the symmetry of the germanium crystal, shown in the figure, which accounts for its excellent conductivity and outstanding interaction with other materials to form the desired semiconductor properties.

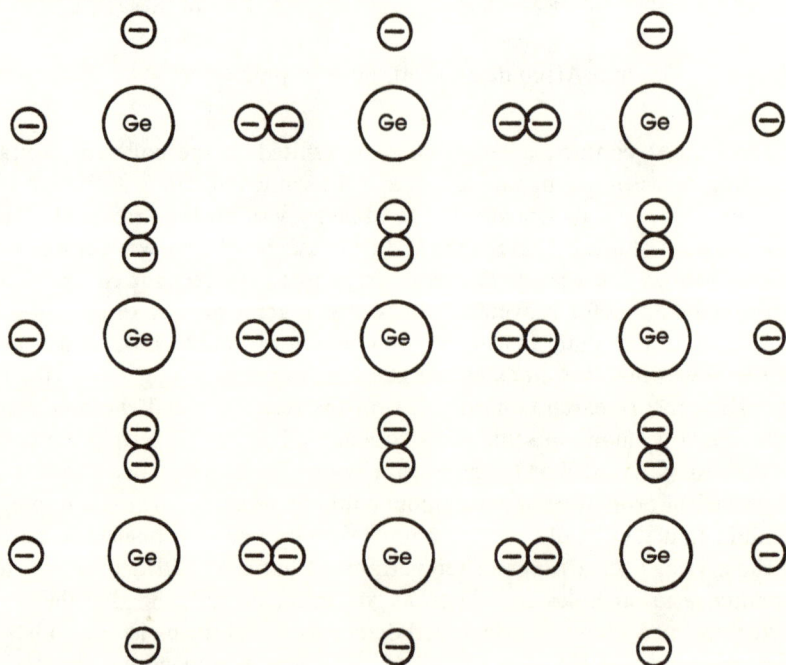

Figure 2.5 Germanium Crystal Structure

The primary use of germanium has been in diodes and transistors. Germanium has always been the best material for achieving the highest maximum frequency of oscillation, f_{max}, in semiconductors. The drawback was that when these devices were fabricated, the base and emitter stripes (pads) were not large enough to allow attachment of bonding wires. Some planarization techniques eliminated some of this problem, so that higher frequency germanium transistors could be fabricated if the need was sufficient.

An important part of semiconductor operation is the energy band within a particular element. Figure 2.6 shows a diagram of the two upper bands of energy within a material. There are other bands below the valence bond band, but for considering the electrical properties of a material the two upper bands are the only ones of interest. Thus these are the ones we will concentrate on.

The *conduction band*, in Figure 2.6, is a band of energy in which the level of energies is high enough to allow electrons at this level to move freely under the influence of any external field or force. That is to say, the electrons will sustain a current rather easily if a voltage is applied.

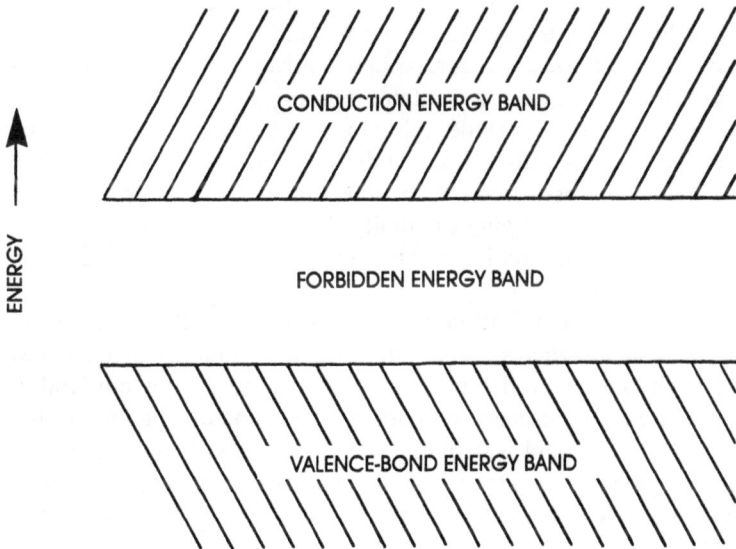

Figure 2.6 Energy Bands

The *valence band* is a band of energies in which the energy is of the same level as that of the valence electrons. This means that the electrons which occupy this level are not free to move around as are those in the conduction band. These electrons, however, may be raised to the conduction band if sufficient energy is applied to the material. To reach the conduction band they must cross the gap between the valence band and the conduction band. This gap is called the *forbidden energy band.* It is the energy difference across this region that determines whether a material is a conductor, an insulator, or a semiconductor (such as we need for our application).

Germanium has the narrowest width of forbidden band that can be used effectively in the construction of solid-state devices (diodes or transistors). This width is 0.7 V and limits the maximum temperature to which the material may be used as a semiconductor. For this reason, germanium (and other materials) must be doped with impurities so that the conductivity by means of positive or negative charges (holes or electrons) will be larger than the conductivity within the material due to thermal effects. This thermal effect is not noticeable at a fixed temperature or over a small temperature range. When a large range of temperatures is required, however, as is the case in most practical applications, this phenomena must be considered. This is one reason why germanium does not find many applications in today's high technology, wide temperature extreme market.

An example of the use of germanium in microwaves is the tunnel diode. (This diode has been used for oscillator and amplifier applications prior to the maturing of gallium arsenide technology.) Although it does not produce the lowest noise figure tunnel diode, it does produce one which will provide a higher gain amplifier circuit than other materials. This, once again, is because of its current carrying capability. It will carry more current which will result in more gain, but it also has a higher noise characteristic because of the increased current.

Other than the application above (tunnel diode), there are not many areas of microwaves where germanium is found in modern-day microwaves. Because of the narrow width of its forbidden band it does not lend itself very effectively to the microwave world which is one of small wavelengths, minute transit times, and wide temperature extremes. In spite of its drawbacks for microwaves, however, we must not discount what germanium has done for the entire semiconductor industry of today — it is the originator of semiconductors.

2.2 SILICON (Si)

Silicon follows only oxygen as the second most abundant element in the earth's crust. It occurs natively as silica in sand and quartz and as metal silicates in most rocks. The use of silicon in diodes and transistors has become extensive and has, in many cases, completely replaced germanium for these applications.

The silicon we are concerned with for microwave applications is *pure silicon*. The methods of preparation of very high purity silicon (>99%) may vary from manufacturer to manufacturer. Also, most of the processes and their details are proprietary and are not disclosed. There are, however, general steps which are taken by all manufacturers which can be told. The first step is the conversion of metallurgical silicon (that silicon which is 96 to 99% pure, made by the reduction of silicon with coke at 1770 K in an electric furnace) either to silicon tetrachloride or, more commonly, trichloroethylene. The silicon halide is purified by fractional distillation and then reduced back to silicon by heating with hydrogen at 1300 K (1027°C or 1880°F), either in a quartz tube or over a heated rod of pure silicon on which deposition of the reduced silicon takes place, thereby increasing the diameter as the process goes on. Trichloroethylene is preferred to silicon tetrachloride at this stage because it is easier to remove phosphorus and boron compounds by distillation and it is more rapidly reduced.

The *pure* silicon obtained is polycrystalline and >99.0%. Most semiconductor devices require single-crystal, ultra-pure silicon with impurity levels of less than one part in 10^9. These levels of impurity are determined by activation analysis, or mass spectrometry, or are empirically determined by measurement of resistivity and other electrical tests. The silicon obtained from reduction of the halide is further purified by the zone refining process in which narrow bands (or zones) are melted in a vertical silicon rod and made to travel repeatedly in the same direction along the length. The molten zone retains its integrity by surface tension. Impurities preferentially dissolve in the molten zone and are carried with it to the end where they are concentrated, leaving the rest of the rod in the ultra-pure form. Extremely small controlled additions of the required additives to give *p* or *n* type characteristics (as described earlier) are then added to the silicon. This doping may be of acceptor atoms (boron or gallium) to produce *p*-type material or of donor atoms (arsenic or antimony) to produce *n*-type material.

After doping, a single crystal of silicon is grown from a small crystal seed by very slowly pulling the growing crystal from the melt. The doped silicon in the single crystal form is then sliced into very thin discs. This is the point where most people perceive the semiconductor process as beginning. It can be seen, however, that the actual process has begun many steps back. After the discs are sliced, subsequent stages of actual device manufacture include the construction of *pn* junctions (usually by vapor deposition of a layer of silicon with dope of the opposite type), division of the disc into tiny chips, selective etching, attachment of connecting leads, and final encapsulation. At this point you have a working silicon-based semiconductor.

Probably the greatest advantage of silicon over germanium is its gap energy or forbidden band energy. In the case of silicon it is 1.1 V as opposed to 0.7 V in germanium. This comparison is shown in Table 2.1.

Table 2.1

Semiconductor	E_g (electron volts)	U_e	U_h
Ge	0.7	3900	1900
Si	1.1	1500	500

As was previously mentioned, a large energy gap (or forbidden band) is useful in semiconductor devices, since a large gap gives a small level of thermal conduction. You will recall that this was one of the disadvantages of germanium. It had the narrowest energy band of all the semiconductor materials. This gap is important when considering materials for semiconductor devices, since you only want conduction through the material because of bias voltages applied to it. In this way you can control the device to meet your particular needs. If you have a large level of conduction through the device due to the temperature the device is subject to, you will never know how that device will act, since you cannot possibly have complete control over it. Thus a wider energy gap is very beneficial for semiconductor material.

The last two columns of the table presented above are indications of charge mobility. They are the electron mobility, U_e, and the hole mobility U_h, within the material. Units of this mobility are cm^2/V-s. Of more importance to the operation of semiconductor devices is their ratio. Since germanium and silicon devices are characterized as *bipolar* devices, that is they rely on both electron and hole movement (or mobility) for their operation, characterizing the relationship of these two charges is a necessity. It can be seen that in germanium there is approximately a 2:1 ratio of electron to hole mobility. This is good, since the electrons are designated as the primary carriers. Silicon, on the other hand, has a 3:1 ratio of electron to hole mobility. This tells you that its primary carrier conduction is better than that of germanium.

Another advantage of silicon over germanium is its higher avalanche or breakdown voltage. This higher voltage capability is due to a large extent to the high inherent resistivity available within silicon. As a comparison, consider the fact that germanium diodes will exhibit a breakdown voltage in the order of 150 to 200 V. The same diode constructed in silicon will have a breakdown voltage of 1000 V.

A major disadvantage that is always quoted when speaking of silicon is its upper frequency limit. This is due to the inherent characteristic of a relatively low value of the diffusion constant for charge carriers. This, as mentioned, is a characteristic of silicon and must be worked around to obtain the other desirable features of silicon devices.

Silicon is used in microwaves for diodes (Schottky, PIN, IMPATT, and TRAPATT) and in bipolar transistors. The only limitation which is encountered is when using the bipolar transistor. Best performance as far as both power and noise performance is generally below 4 GHz. Devices designed and used in this range will generally perform well. With this limitation kept in mind and proper design procedures used, the silicon devices can be used very effectively for a variety of microwave applications.

2.3   GALLIUM ARSENIDE (GaAs)

Gallium arsenide, as the name implies, is a compound. It consists of a group 3 element (gallium) and a group 5 element (arsenic). Notice that we are now talking about a material that has two elements precisely joined as opposed to the manufacturing of a pure element such as silicon. The first element, gallium, is a silvery metal with a melting point just slightly less than the surface temperature of the human body (29.8°C or 85.6°F). It is one of only four metals that can liquify at or near room temperature. The others are mercury (–38.87°C or –37.96°F), cesium (28.5°C or 83.3°F) and rubidium (38.5°C or 101.3°F). Gallium was discovered by Paul E. Lecoqde Boisbaudran in 1875 through the use of the spectroscope and electrolysis. The name *gallium* is derived from the Latin *Gallia*, which means France.

The second element in gallium arsenide is arsenic. This element is a dark gray semimetallic solid which can be traced back to around the year 1250 when it was discovered by Albertus Magnus. From its earliest beginnings it has been known as a lethal poison. If you mention the word *arsenic*, you will probably see the same association made by those who hear the term. Its name is derived from the Greek word *arsenikon*, which means "yellow orpiment." This element, which sounds so brilliant and lustrous, is at the present time classified by the Environmental Protection Agency as a suspected carcinogen. That is, it is suspected of causing cancer under certain circumstances.

If you think of these two elements by themselves, you would have a difficult time ever believing that when they were combined they would form a compound that would allow the field of microwaves to make one of the largest advancements in its history. This combination has ruled the field from the very time it first appeared in the early 1970s. Although it sounds like a very lethal combination, the material, GaAs, a grayish and brittle semi-conductor, is relatively benign chemically and is safe to handle under normal conditions. Gallium arsenide is characterized as a direct-band-gap semiconductor with an isotropic minimum at the center of the Brillouin zone. This is a zone named after Louis Brillouin for his contributions to the understanding of wave motion in periodic media. The zone is when the wave number, k, is equal to the range between $-\pi/a$ to $+\pi/a$. Figure 2.7 shows a comparison of energy levels for this zone for silicon (Figure 2.7a) and gallium arsenide (Figure 2.7b). It can be seen that there is a definite difference in the area of the forbidden zone. The silicon zone is much smaller. Also, the electron energy (mobility) is much greater in the gallium arsenide. This makes the material an excellent choice for high frequency operation. If we go back to the table presented in the previous section and add gallium arsenide to it we can see these two points brought out vividly.

Table 2.2

Semiconductor	E_g (electron volts)	U_e	U_h
Ge	0.7	3900	1900
Si	1.1	1500	500
GaAs	1.43	5000	400

It can first be seen that GaAs has a higher gap energy (1.43 eV as opposed to 1.1 for silicon and 0.7 for germanium). This allows the devices to operate at higher temperature than either of its two predecessors.

Another point to be emphasized is the higher mobility (cm^2/V-s) of the electron in particular within the GaAs material. It can be seen that the ratio of electron mobility to hole mobility is 12.5:1 as opposed to 2 or 3:1 for germanium and silicon, respectively. This is necessary for high-frequency operation, since the GaAs devices are *unipolar* devices, as opposed to *bipolar*, which the germanium and silicon devices are. The term unipolar means that the primary means of conduction is by electrons. Thus the ratio of electrons to hole mobility must be high. It can be seen that this is true for the GaAs material.

The combination of a large energy gap and high electron mobility makes gallium arsenide an ideal material for high frequency, high temperature, and radiation-resistant devices. In contrast, however, to the growth of bulk crystals from a monatomic and low vapor pressure material such as silicon, the synthesis of gallium arsenide is greatly complicated by the previously discussed volatile and toxic element, arsenic. By using special handling procedures and special growth techniques, very high quality single crystals can be grown and processed into high performance devices.

(a) SILICON

(b) GALLIUM ARSENIDE

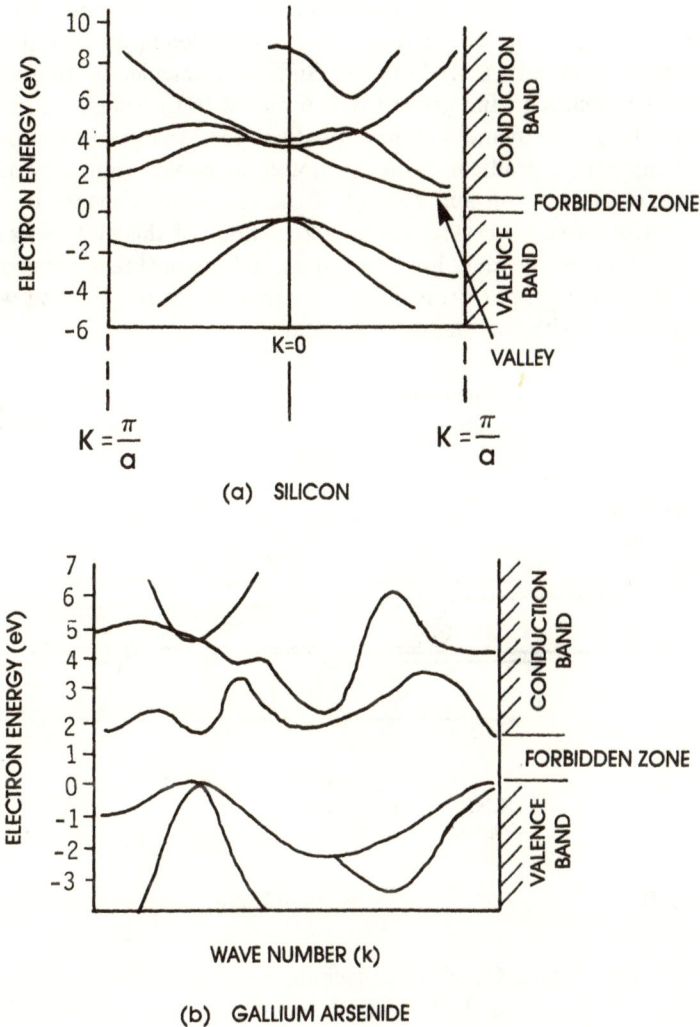

Figure 2.7 Brillouin Zones

During the course of crystal growth, gallium and arsenic are first compounded in an exothermic (liberation of heat) reaction. Single crystal GaAs is then grown from the molten material which results. Substrate materials are then sliced and polished from the single crystals. The major commercial GaAs crystal growth methods fall into two categories. They are the horizontal Bridgeman technique and the vertical Czochralski technique.

Figure 2.8 shows the Bridgemen horizontal technique. In this setup a GaAs charge is formed separately or compounded in situ in a quartz crucible or PBN boat with a seed crystal. Sufficient arsenic to maintain adequate vapor pressure during growth is also placed in the sealed quartz tube. The crystal is grown by moving the entire furnace so that the tube moves through a temperature gradient from the hotter to cooler section of the furnace. Careful observation and control of the seed-melt interface is required for initiation of single-crystal growth. Variations of this system have included use of four independent hot zones and a sodium heat pipe for precise control of the arsenic vapor pressure and elimination of the viewing window and replacing it with thermal monitoring devices.

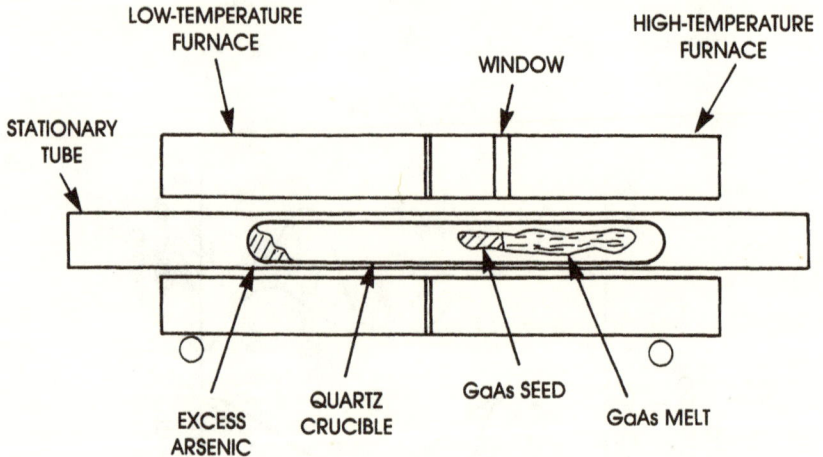

Figure 2.8 Bridgeman Horizontal Technique

The vertical Czochralski technique is referred to as the liquid-encapsulated Czochralski (LEC) growth technique. This process is illustrated in Figure 2.9. In this technique the vaporization of arsenic (As) from molten GaAs is inhibited by placing a layer of nonreactive boric oxide (B_2O_3) on the melt surface. An inert gas pressure, which is higher than its arsenic partial pressure, is then maintained on top of the B_2O_3 encapsulating layer, and growth proceeds as it would in a standard Czochralski technique. This standard process is as follows:

A seed crystal is dipped into a melt whose temperature is lowered until a small amount of crystalline material is solidified. The crystal is then slowly withdrawn from the melt at a typical rate of 1 to 10 mm per hour. The temperature of the melt is carefully adjusted to produce a crystal of the desired diameter.

Other methods that may be incorporated for growing of GaAs are *horizontal gradient freeze*, *zone melting*, liquid-encapsulated Kyropoulus (LEK), Magnetic LEC (MLEC), and a variety of techniques tailored to individual manufacturers. The best method to use is the one which will provide the purest and most consistent GaAs crystals. Generally, the method is either the LEC or MLEC method.

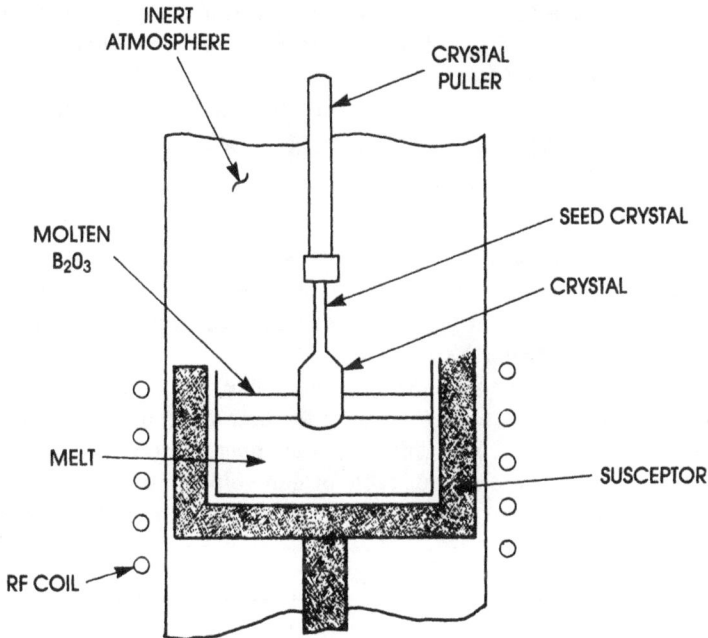

Figure 2.9 LEC Technique

2.4 INDIUM PHOSPHIDE (InP)

The latest compound to make its presence known on the microwave scene is indium phosphide (InP). This material has made solid state devices operating efficiently up into the hundred gigahertz region very practical and attainable. The compound, obviously, is made up of two elements: indium and phosphorous.

Indium is a soft, silvery, malleable metal which has found applications in microwaves as an excellent low temperature solder. It was discovered in 1863 by two German scientists, Ferdinand Reich and Hieronymus T. Richta, and received its name because of its indigo blue spectrum. The metal is widely distributed in nature but occurs in commercially practical concentrations only in ores of zinc, iron, lead, and copper. Because of this it did not have many applications and for many years was only a laboratory element. It was not until 1924 that William S. Murray decided to do something about the situation and take indium out of the laboratory and make it work commercially. In 1924 Murray received a patent for his indium electroplating process. In 1934 he founded the Indium Corporation of America in Utica, New York. This was the world's first commercial producer of indium metal. From this point indium has become a very important metal for a variety of microwave applications, not the least of which is forming the compound InP for solid-state applications.

Phosphorous is a nonmetallic element that appears in three varieties:

1. a yellow, which is very reactive, soft, and poisonous,
2. a red, which is less reactive and nonpoisonous,
3. a black phosphorous.

The element was discovered in 1669 by Hennig Brad by heating white sand with evaporated urine. Its name is derived from the Greek *phosphorus*, which means "light-bearing." It is also classified as the eleventh most common element in the earth's crust.

Once again, as with GaAs, it seems like an unlikely combination to put these two elements together, but their combination results in a material which shows excellent performance in the upper microwave and the millimeter bands. If we complete our table of gap voltage and mobility parameter by adding indium phosphate to it, we can see some interesting comparisons:

Table 2.3

Semiconductor	E_g (electron volts)	U_e	U_h
Ge	0.7	3900	1900
Si	1.1	1500	500
GaAs	1.43	5000	400
InP	1.25	4000	100

One area to note is that the gap voltage is not as high as GaAs, but not as low as silicon. This indicates that indium phosphide will not show as good a performance in temperature as GaAs. The large advantage, however, can be seen in the mobility numbers (U_e and U_h). There is a ratio of 40:1 of electron to hole mobility. This accounts for the excellent higher frequency operation.

When we compare InP to GaAs, there is only one area where GaAs is better for millimeter and higher microwave application. That area is mobility, as shown in Table 2.3. All other areas are dominated by InP. The shorter time constant (0.75 ps as compared to 1.5 ps for GaAs), related to the electron energy transfer effect, allows the InP Gunn devices to operate effectively in their fundamental mode through 100 GHz. These devices will be discussed further in Chapter 4 of this text. The higher operating velocities (*saturated velocity* and *effective transit velocity*) in InP provide dimensional advantages in the actual device lengths. These velocities are shown in Table 2.4:

Table 2.4

Velocity	GaAs	InP
Saturated Velocity (cm/s at 500°k)	5×10^6	6×10^6
Effective Transit Velocity (cm/s)	0.7×10^7	1.2×10^7

The advantages of indium phosphide (InP) are numerous and will be expanded upon as applications are presented throughout the text. At this point it is sufficient to say that for application from 20 GHz to 150 GHz there is no better material to use than indium phosphide.

2.5 SUMMARY

Four of the materials used for microwave solid-state devices have been presented in this chapter. The properties, advantages, disadvantages, and history of germanium, silicon, gallium arsenide, and indium phosphide have been shown. It should be clear by this time that each material has its specific application and desirable frequency range of operation. The purpose of presenting all of them is to introduce the critical materials used in microwave solid-state devices so that presentations made further on in the text will be more understandable. It is hoped that this chapter will make the readers much more knowledgeable about the devices they are using.

Chapter 3
Solid-State Junctions

The heart of any solid-state device, whether it is a microwave device or a simple dc switching transistor, is the junction (or combination of junctions) within that device. It makes no difference if we are dealing with a two-element (diode) or three-element (transistor) device, they all consist of junctions of material of one form or another. It is the characteristics of these junctions that make each device unique and different from all others.

If you look up the word *junction* in the dictionary you will find that it is: *the line or point at which two bodies are joined.* This definition very accurately defines a solid-state junction, since we are concerned with what takes place at the point where the two semiconductor materials come together and not necessarily with what the individual materials are doing by themselves. It is, in actuality, the operation of the two materials *after* they are brought in contact with one another that results in the desired operations. That is, the characteristic of the junction given the overall operation of the device. Nowhere is this junction operation more important than in solid-state microwave devices. Time and distances are of the utmost importance when dealing with devices which will have the capability of operating from 1 GHz upwards into the 100 GHz ranges. Thus knowing how a semiconductor junction operates in a quiescent state, and when forward and reverse biased is of paramount importance in understanding how the various diode and transistor configurations perform and how they can do the best job for your application.

We will cover two types of solid state junctions in this chapter. They are the PN junction and the Schottky junction. Each of these junctions will be presented; theory of operation will be outlined in detail; and their operation under both forward and reverse biased conditions will be shown. Understanding of these junctions will aid greatly in being knowledgeable of microwave solid-state devices.

3.1 PN JUNCTIONS

When most people think of a junction in a semiconductor device, the PN junction is the one which most commonly is mentioned. This is the type of junction that was used by Karl Lark-Horovitz at Purdue in 1942 when he did extensive work with germanium for use in rectifiers; it is also the type of junction used by Brattain and Bardeen at Bell Telephone Laboratories in 1948 when they demonstrated the first transistor, a point contact device; and it is still the junction used for many microwave diodes and bipolar transistors that are available today.

A PN junction is formed when a *p*-type material and an *n*-type material are placed in physical contact with one another. You will recall from our discussion in Chapter 2 that a *p*-type material is formed when an atom from the third column (boron, for example) is added to a germanium or silicon atom (Column IV), and there are too few electrons available to make a covalent bond. This causes a void, or *hole*, which constitutes a condition termed an *acceptor* atom. In this condition the element formed will accept an electron if one is available. This is now *p*-type material with the *p* representing a positive charge — the charge of the vacated hole. Similarly, an *n*-type material is formed when an atom from the fifth column (arsenic, for example) is added to a germanium or silicon atom. There would be covalent bonding, but there would also be an excess electron as a result of this combination. This is because arsenic has five electrons in its outermost orbit as opposed to only four for either germanium or silicon. This extra electron is termed a *free electron* and results in a *donor* material which is classified as *n*-type. The *n* in this term is a designation for *negative*, which is the charge of the electron, which is the primary current carrier in this case.

Thus, the *p*-type material (acceptor, positive charge, holes being the primary current carriers) and the *n*-type material (donor, negative charge, electrons being the primary current carriers) form a junction when they are physically placed together which allows an operator to conduct a variety of control functions depending on a specific application. The operation of this PN combination is what will be investigated in detail.

As stated previously, a PN junction is formed when a *p*-type material and an *n*-type material are joined (bonded) together. Immediately after formation, carrier diffusion results in some electrons from the *n* region crossing the boundary, while some holes from the *p*-region migrate to the other side. After crossing the junction, an electron which originated in the *n* region is in a high hole concentration environment. In this situation recombination is highly probable, and the electron as a carrier is thus annihilated. Similarly, when a hole crosses the junction from the *p* region, it encounters a high electron concentration, recombines, and is annihilated as a carrier. Since each region was originally electrically neutral, electron-hole recombinations on both sides and in the vicinity of the junction result in layers of ionized acceptors in the *p*-region and donors in the *n*-region. It can be said that the *p*-region has experienced a net accumulation of negative charge buildup continuous until an equilibrium condition occurs and further carrier diffusion across the junction is negated by the repelling force between the carrier and charge concentration across the boundary. The PN junction in an equilibrium state is shown in Figure 3.1. The "built-in" electric field produced at the junction in an equilibrium condition is of such a direction and magnitude that, as previously stated, it opposes the diffusion of any mobile carriers across the junction. Thus, with no bias applied to the formed crystal, there is no net flow current. A dynamic equilibrium therefore exists within crystal in which a small current of minority carriers diffuses to the junction and is drawn across by the built-in field referred to earlier. Similarly, an equal current of majority carriers diffuses across the junction in the opposite direction. The potential difference within the crystal resulting from the built-in field is called the *barrier potential*. In most PN junctions this is in the order of 0.6 V.

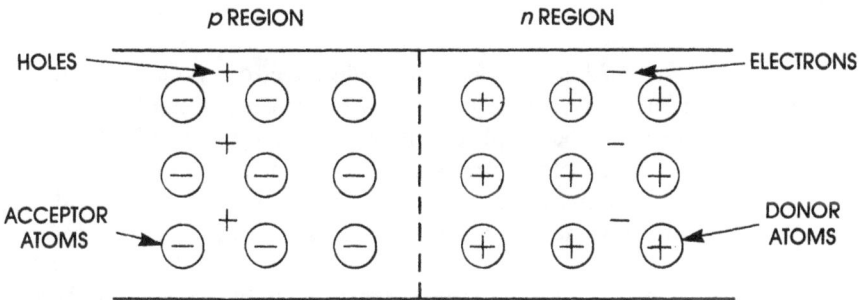

Figure 3.1 PN Junction at Equilibrium

An important aspect of the PN junction are the energy levels involved as the two materials are joined to form the junction. Figure 3.2 shows the energy level diagram for a PN junction. Notice that the *p* region is at a higher energy level than the *n* region. This is because the *p*-type material has lost some high energy holes and gained some high energy electrons, as described earlier. This section is no longer neutral at this point and thus will be displaced above the energy level of the *n*-type material by an amount equal to qV_j. In this quantity, q is the charge of the electron and V_j is the electrostatic potential difference across the junction. For an electron to move from the *n* region to the *p* region, it needs to have sufficient energy to climb

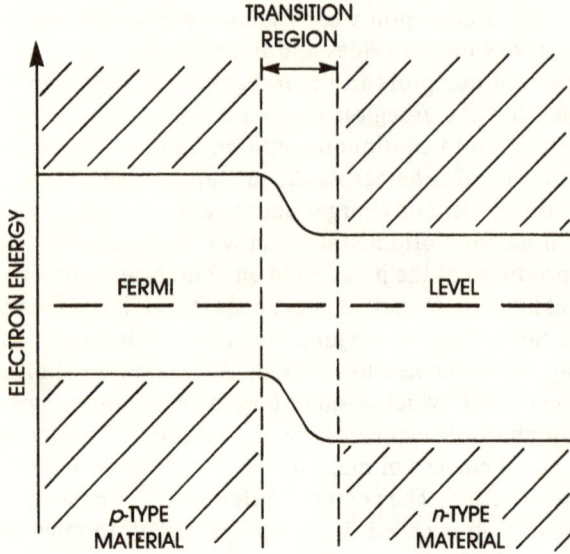

Figure 3.2 Energy Level Diagram

the resulting *potential hill* (the electron would have to invade the *p* region which has more negative charge than the *n* region). Also, the holes in the *p* region would need to be supplied with external energy in order to climb the potential hill in the other direction. Thermal agitation accounts for a small amount of carrier flow across the junction, but this is minimal. In general, all of the movement of majority carriers is completely balanced by minority carriers.

One term that is shown on the energy diagram in Figure 3.2 is the *Fermi level*. Very basically put it is a reference energy level at the PN junction. There is, however, much more to defining a Fermi level than saying it is a reference level. The filling of energy bands by electrons is accomplished by a simple rule: states of lowest energy are filled first, the next lowest level, and so on. Finally, all of the electrons have been accommodated. The energy of the highest filled state is the *Fermi level* or *Fermi energy* (W_f). The magnitude of W_f depends on the number of electrons per unit volume in the material. The expression which is used in conjunction with the Fermi level is called the *Fermi distribution function* and expresses the statistical probability that an electronic state of energy W in a crystal is occupied by an electron at a given instant. The expression is:

$$F(W,T) = \frac{1}{1+e^{(W-W_f/kT)}}$$

where:

        W_f = Fermi level
        k = Boltzmann's constant (1.380×10^{-23} J/K)
        T = Absolute temperature

A plot of the function is shown in Figure 3.3. Notice that this curve is symmetric about the midpoint. When the energy W is less than W_f, the function $F(W,T)$ is approaching unity and when W is greater than W_f, the function is approaching zero. The relationship of this function to a semiconductor is shown in Figure 3.4. In this representation the Fermi level is located in the center of the energy gap so that the number of electrons in the conduction band is equal to the number of holes in the valence band. This is an *intrinsic* semiconductor. The intrinsic semiconductor is defined as:

A material whose conductivity lies between those of insulator and conductors; its valence is full, and is separated from the conduction band by an energy gap small enough to be surmounted by thermal agitation; current carriers are electrons in the conduction band and holes in the valance band in equal amounts.

If this definition sounds familiar, it should. It is the definition for a perfect, undoped semiconductor. These were referred to in Chapter 2, when discussions of germanium and silicon were presented.

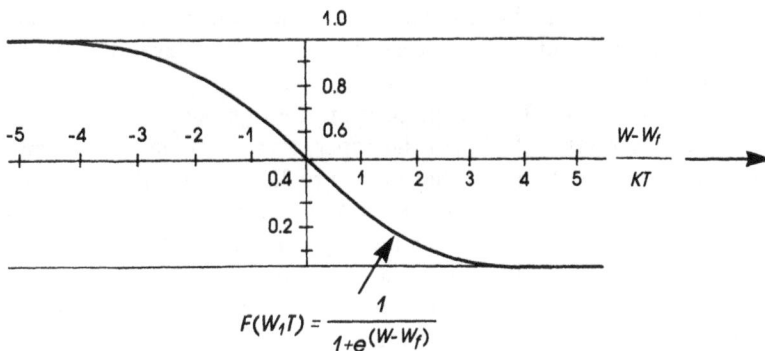

Figure 3.3 Fermi Distribution Function

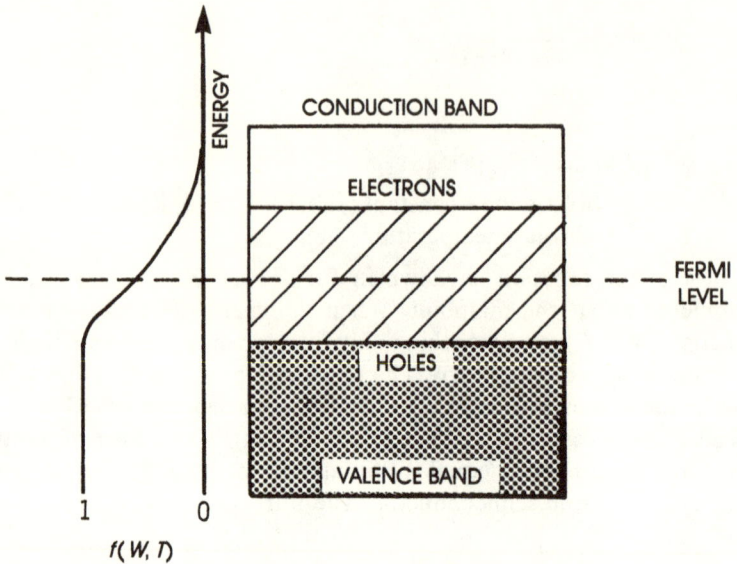

Figure 3.4 Fermi Distribution in a Semiconductor

In an *n*-type semiconductor (heavily doped with donor atoms), the electrons from ionized donors can fill the holes in the valence band as well as enter the conduction band. This shifts the whole distribution upward, as shown in Figure 3.5. Since the donor concentration increases, the Fermi level is raised, as shown.

Conversely, in a *p*-type semiconductor (heavily doped with acceptor atoms), the acceptor band can accommodate electrons from the conduction band and valence band. This process shifts the Fermi level downward, as shown in Figure 3.6.

If we take Figure 3.5 and place it on the right side of the paper and place Figure 3.6 on the left side of the paper, we see a remarkable similarity between this combination and that of Figure 3.2. This similarity should be there, because this is actually what the PN junction semiconductor consists of — a *p*-type material joined with an *n*-type material. It can also be seen that the Fermi level is shifted upward on the *n* side and downward on the *p* side, just as described in the individual doping explanations. Thus it can be seen how the energy levels of a PN junction will vary with doping contents of the individual semiconductor.

Discussions have centered on either the *p* region or the *n* region and how they are combined at a junction. There is, however, an area on both

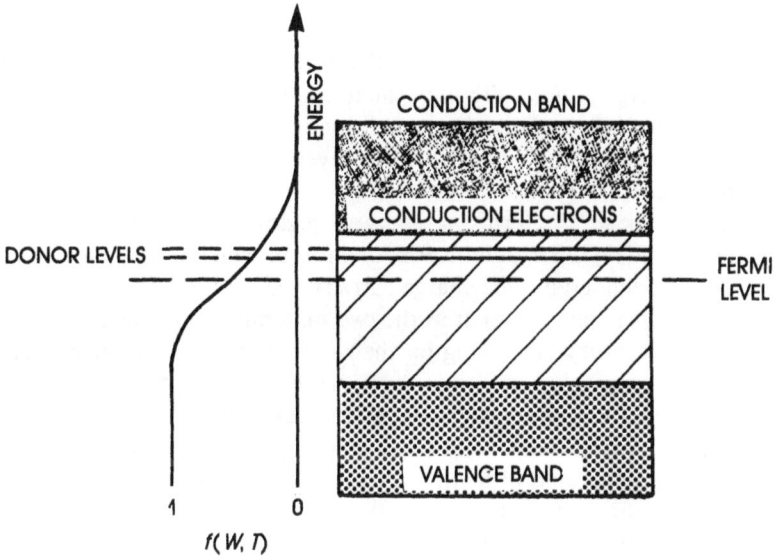

Figure 3.5 *N*-Type Semiconductor Fermi Level

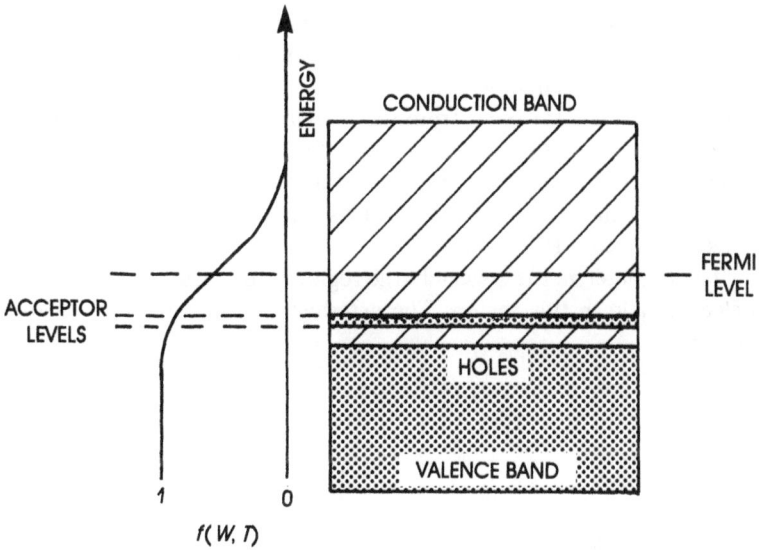

Figure 3.6 *P*-Type Semiconductor Fermi Level

sides of this junction which must also be defined. This area was termed the *transition region* in Figure 3.2. It is also referred to as a *depletion region* or a *space-charge region*, although the term depletion region is used when the PN junction is in a reverse-bias condition rather than a no-bias condition, as we are now considering. (Forward and reverse conditions will be discussed later in this chapter.)

If we look up the term *space-charge* region we will find a definition as follows: The region around a PN junction in which holes and electrons recombine leaving no mobile charge carriers and a net density difference of zero. What this is saying is that as the two materials come together to form the PN junction they are taking on their own individual charge and characteristics. During this *equilibrium* process the carriers in the neighborhood of the junction are swept away from the junction in their appropriate direction. This causes the junction area, and a minute distance on either side of it, to become an area of neutral charge. This can be seen in Figure 3.7. The top portion of the figure shows the PN junction with its appropriate acceptors, donors, holes, and electrons. The junction follows down through to parts (b) and (c) of the figure, and the distance from the junction on the right and left are distances through the semiconductor away from the junction. Figure 3.7(b) shows the charge density at and around the junction. It can be seen, as mentioned previously, that the net charge in the space-charge region is zero. Aso shown is a *potential hill* in Figure 3.7(c). This is an electrostatic potential which is present as you progress from the junctions into the semiconductor material. The one shown in solid is a formidable obstacle for holes from the *p* region to overcome. Similarly, the dashed line also shows that there is a hill of equal magnitude which presents a large obstacle for the electrons from the *n* region. These potential hills further substantiate the neutrality of the space-charge region, when there is no bias applied to a PN junction.

To characterize the space-charge region you must calculate the potential electric field in this region. The net charge in the area can be approximated by

$$\rho = e(N_d - N_a)$$

where:

e = charge of the electron
N_d = density of the ionized donor atoms
N_a = density of the ionized acceptor atoms

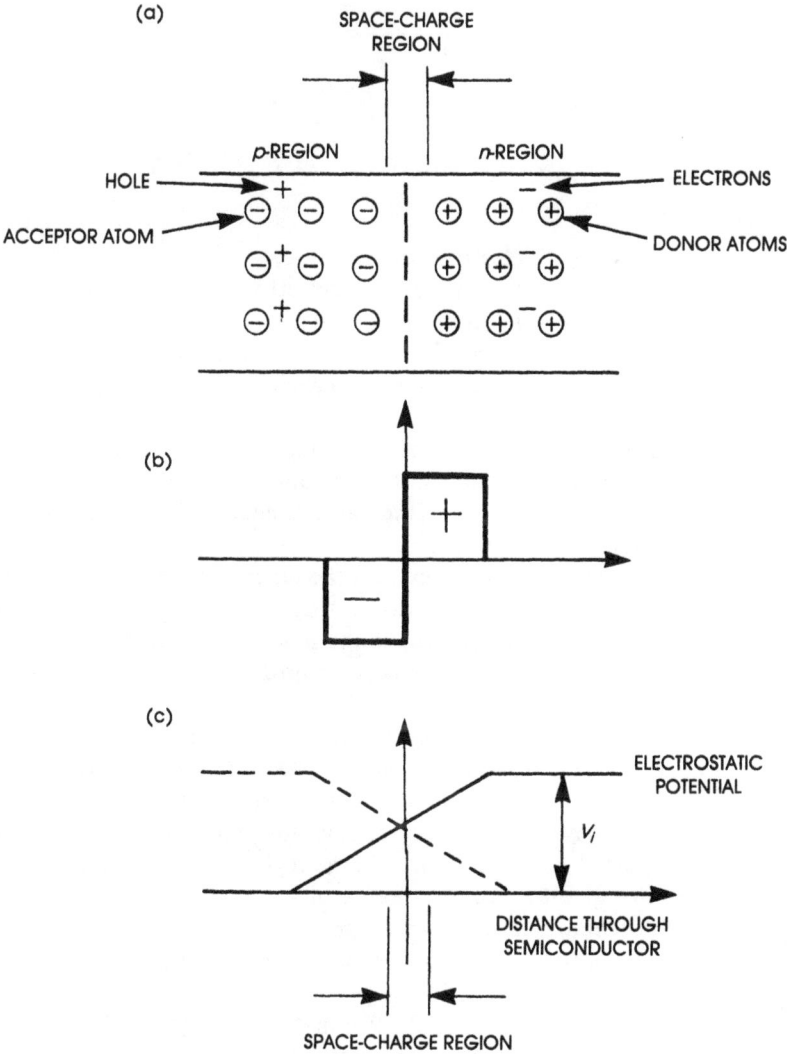

Figure 3.7 Space-Charge Characteristics

This equation is called the *space-charge approximation*. The charge density of the region actually, however, depends on the impurity distribution within the space-charge region. This can be achieved as follows:

1.   Write Poisson's equation as

$$\nabla^2 \psi = \frac{-e \ N(x)}{\epsilon}$$

where

$$
\begin{aligned}
\psi &= \text{electrostatic potential;} \\
-e &= \text{the charge of the electron;} \\
N(x) &= \text{net impurity density;} \\
\epsilon &= \text{permittivity (dielectric constant) of the} \\
&\quad \text{semiconductor material.}
\end{aligned}
$$

2.   Integrate the equation twice to obtain ψ. When you achieve the two integration constants, they are obtained by requiring that:
      a. The electric field vanishes at the edges of the space-charge region;
      b. the potential is continuous at the junction.
3.   The boundary conditions are $\psi = \psi_n$ and $\psi = \psi_p$ at the edges of the space-charge region (ψ_n is the potential in the bulk n region; ψ_p is the potential in the bulk p region).

As a result of the diffusion process within the region of the PN junction, a potential gradient is established across the space-charge region. This gradient is represented by an imaginary battery, as shown in Figure 3.8. The battery is termed *imaginary* because it is only shown to represent the internal effects of the junction and the space-charge region. The actual potential is not a measurable quantity. In the absence of any external voltages (no bias applied), the potential gradient discourages further diffusion across the PN junction because electrons from the n-type material are repelled by the very slight negative charge induced in the p-type material. Thus, the gradient prevents total interaction between the two materials (n-type and p-type) and consequently preserves the differences in their individual characteristics.

Two junctions which can be analyzed as discussed above are the *abrupt* junction and the *graded* junction. The abrupt junction is one in which the impurity concentration (assuming that the impurities are fully ionized) is constant on both sides of the junction. In Figure 3.9 the charge density for

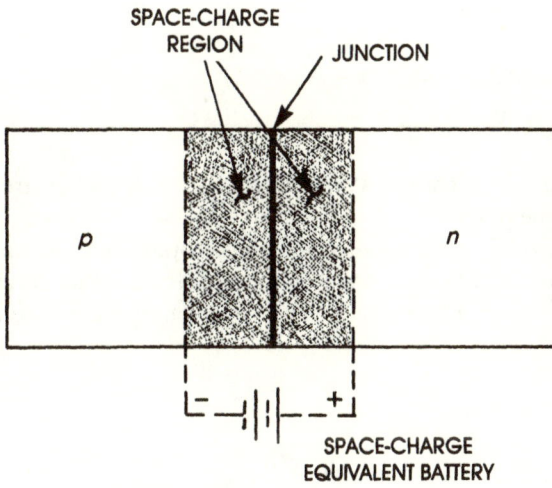

Figure 3.8 Imaginary Space-Charge Battery

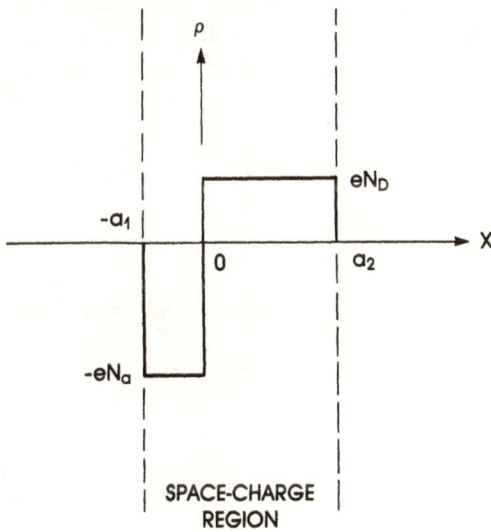

Figure 3.9 Abrupt Junction

this junction is given by

$$\rho = \begin{cases} -e \, N_a & -a_1 < x < 0 \\ e \, N_d & 0 < x < a_2 \end{cases}$$

where the origin is chosen at the junction and a_1 and a_2 are boundaries of the space-charge region.

The *linearly graded junction,* on the other hand, is one in which the impurity concentration is a linear function of the distance across the junction. This is shown in Figure 3.10. The space-charge approximation for this junction is

$$\rho = eN(x) = egx$$

where g is the impurity *gradient.*

The space-charge approximations described in the text above allow you to calculate the width, w, of the space-charge region. Knowing this width makes it possible to determine what the junction capacitance of the PN junction is. This is of prime importance when dealing with higher frequency devices such as those used for microwave applications. Junction capacitance will be covered in detail following the discussions on biased PN junctions. This is because the capacitance has much more meaning when the junction is either forward or reverse biased.

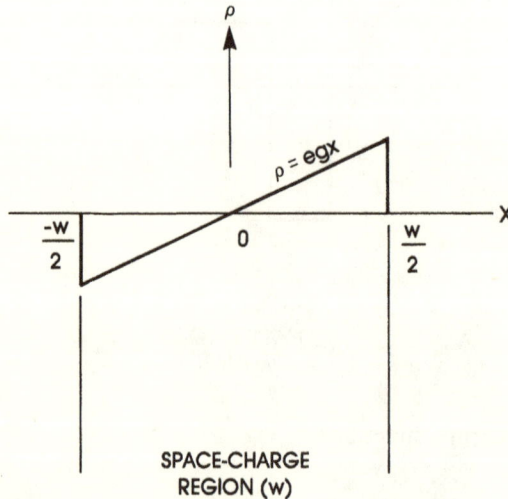

Figure 3.10 Graded Junction

All of our discussions on PN junction have thus far had everything in an equilibrium state. That is, one side has holes and acceptor atoms (*p*-region) while the other has an equal number of electron and donor atoms (*n*-region). There were no external voltages applied to disturb this state. To understand the operation of the junction, however, we must apply voltage, since, in actuality, all microwave solid state devices are active devices (require voltage for operation). It is now time to apply a *bias* to the PN junction we have worked so hard to create.

There are two conditions we can achieve by biasing a PN junction. One condition is *forward* biased and the other is *reverse* biased. Both conditions must be understood if an individual is to comprehend the capabilities of the PN junction devices available for circuit applications. For you to gain the proper perspective and grasp the operation of a biased PN junction, we will compare it to the unbiased PN junction we have been discussing above.

You will recall from previous discussions that the unbiased PN junction energy diagram, as shown in Figure 3.2, indicated *p*-type material energy levels slightly higher than those of the *n*-type material. This is shown again in Figure 3.11(a) and is due to the *p*-type material losing some of the high energy holes while gaining high energy electrons. The amount of the difference, as previously stated, is qV_j with q being the charge of the electron and V_j the electrostatic potential difference across the junction. This results in a potential hill which must be overcome when conduction is to be accomplished. In the unbiased PN junction case, however, conduction is not what the designer is looking for primarily, although when the electrons are raised to a sufficient level conduction will occur. This is used when unbiased diodes are used for detection or mixing purposes. In these cases the microwave energy is responsible for allowing the electrons to achieve sufficient energy to cross the junction.

Figure 3.11(b) shows the unbiased PN junctions electrical characteristics. There is a junction capacitance, dependent on the width of the transition region; and electric field which is generated because of the difference in the materials which have been joined; and then there is the equivalent (imaginary) space-charge battery which was described earlier. The importance of these characteristics is *not* any specific value for any of them. The importance is the relative change in each of these characteristics as the junction is biased in a forward or uneven direction.

An important condition to have in a PN junction device is to have it forward biased. In this condition current can flow across the junction very freely. A look at Figure 3.12 will show why this current flows so freely. Figure 3.12(a) shows that when the forward bias is applied the potential hill is dramatically reduced, so that carriers can cross the junction with great ease. Further reduction of the hill by increasing the bias will result in an even

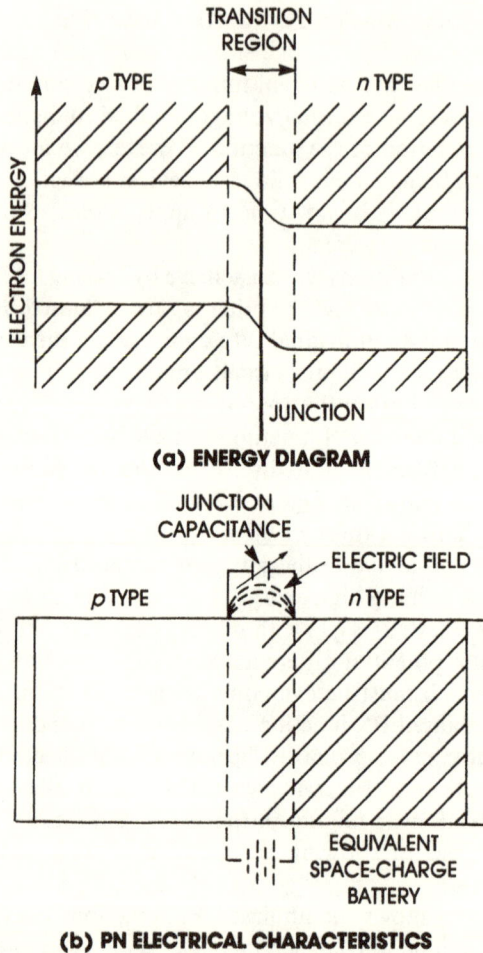

Figure 3.11 Unbiased PN Junction

greater number of carriers crossing the junction. There is, however, a maximum amount of voltage that can be applied. Exceeding this value will destroy the junction and its original characteristics.

Figure 3.12(b) shows how the junction reacts electrically with a forward bias applied. Notice how the transition or space-charge region is smaller than that shown with no bias applied to a PN junction. This is because, with a forward bias applied, the area around the junction has many more carriers in it and is, thus, less of a dead or neutral zone. There obviously is more conduction in this region than with no bias, so naturally the region will be

(a) ENERGY DIAGRAM

(b) ELECTRICAL CHARACTERISTICS

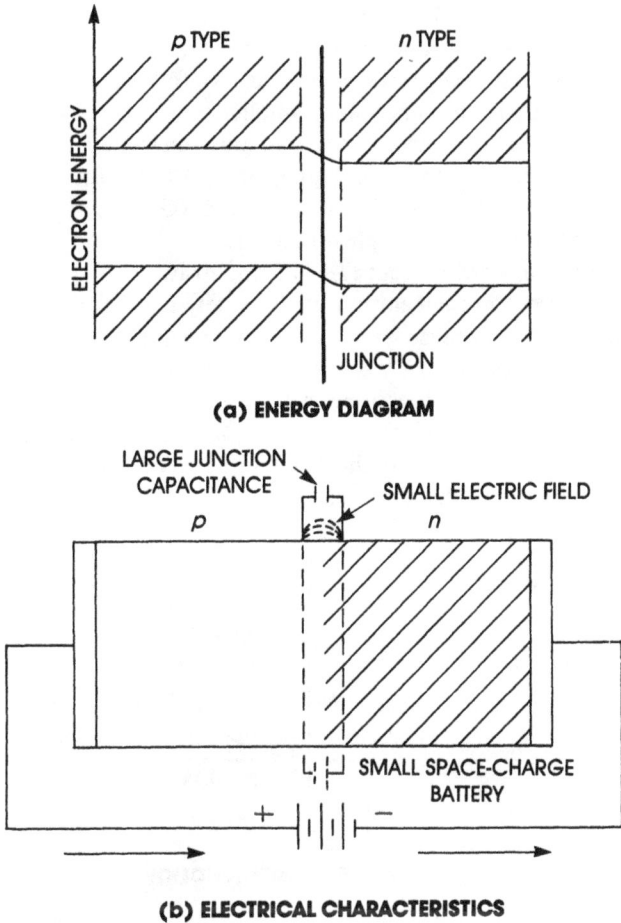

Figure 3.12 Forward-Biased PN Junction

smaller. You will also notice that the junction capacitance is shown to be large as compared to an unbiased junction; the electric field is small; and the space-charge battery is small. These are all a function of the space-charge region.

The junction capacitance is larger because the *effective* plates of a capacitor (the edges of the space-charge region) are closer together. If you refer back to basic ac theory, you will recall that this does increase the capacitance. Similarly, the electric field across the region is smaller, since there is less region across which to develop a field. As a consequence, the space-charge equivalent battery will be smaller. Since the electric field develops across less of a junction (its *width* is decreased) and its intensity or amplitude is less (*height* is decreased), it follows naturally that the equivalent (imaginary) battery will be smaller.

A curve used to characterize a biased junction is the *I-V* curve (current-voltage curve). This is shown in Figure 3.13(a) for a forward-biased PN junction. Curves are shown for both germanium (Ge) and silicon (Si). It can be seen that when voltage is applied in a forward direction, there is a point when very little forward current (I_F) is being drawn. For germanium there is very little current present until approximately 0.3 V is applied and 0.6 V for silicon. This is because the space-charge equivalent battery has an approximate potential of 0.3 V for germanium and 0.6 V for silicon. This battery voltage must be overcome before any current can be drawn. A negative voltage applied to the junction will cause an entirely different reaction. This will be covered in the discussions on reverse bias PN junction. (I_R is reverse current in the junction.)

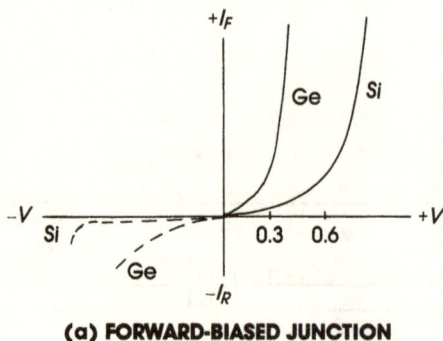

(a) FORWARD-BIASED JUNCTION

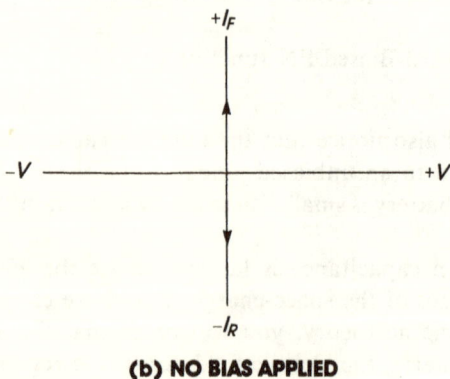

(b) NO BIAS APPLIED

Figure 3.13 *I-V* Curves

Figure 3.13(b) is an *I-V* curve for a PN junction with no bias. Notice how it differs drastically from that of Figure 3.13(a), a forward-biased junction. Since no bias is applied, there is no positive or negative voltage on the device. Thus, the forward current, I_F, equals the reverse current, I_R, and the net result is no current flowing through the device. This verifies what we discussed in earlier sections of this text on unbiased PN junctions.

There are times when zero bias or a forward-biased condition is not what a designer desires. One such time is when a PN junction is being used as a switching diode. To ensure that the diode does not conduct current, a *reverse* bias is applied to the junction. With this bias applied to the junction, Figure 3.14, the potential hill previously discussed which the electrons must climb in order to produce current flow, becomes very high and difficult to climb. This is shown in Figure 3.14(a). The only current possible under these conditions results from a few minority carriers on each side of the junction (electron and hole leakage current). In Figure 3.14(b) you can see how the width of the space-charge region has increased dramatically. This increased width results in a small junction capacity, since the plates of the capacitor have effectively been moved further apart; there is a high value of electric field over this wider area; and the space-charge battery is a larger value, since the electric field intensity is high. This, obviously, has created a much different condition than when the junction was forward biased.

As discussed above, the only current possible under these reverse-biasing conditions is the result of a few minority carriers on each side of the junction. This leakage current is almost completely independent of the magnitude of the reverse-bias voltage until the *avalanche*, or *zener*, voltage is exceeded. This is shown in Figure 3.15. At this point, reverse current (I_R) increases very rapidly. The avalanche breakdown, as it is called, occurs when the electric field across the junction produces ionization as resulting high energy carriers collide with valence electrons. The breakdown appears to be a *field-emission* phenomenon with the strong field in the junction region pulling carriers from their atoms. It should be clear that the junction should not be left in this state for any period of time or severe damage or destruction of the junction can occur.

With all of the biasing conditions covered for the PN junction, it is now time to name the junction we have been discussing. This two-element device is actually a *diode*. That is, it will conduct current in one direction and not conduct it in the other. Extensive coverage of diodes will be presented in Chapter 4. For now we will be concerned with the diode, not as a microwave device, but as a two-element PN junction.

The *I-V* characteristics of the PN junction (or diode) are shown in Figure 3.16. This curve is the result of the following equation:

$$I = I_R \left(e^{qV/kT} - 1 \right)$$

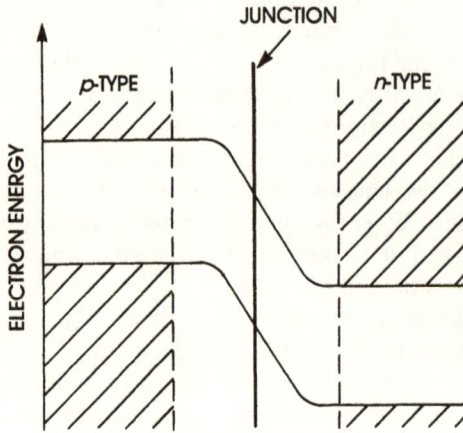

(a) ENERGY DIAGRAM

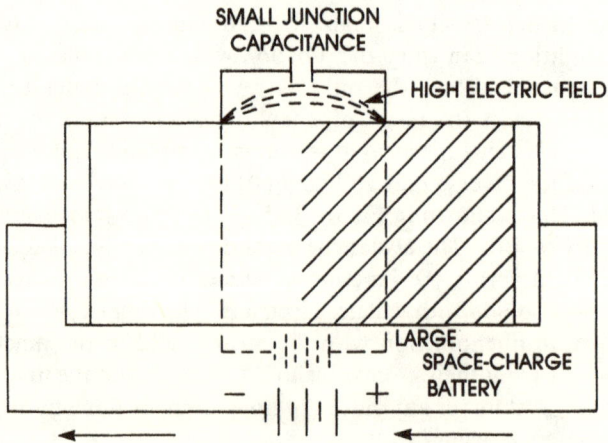

(b) ELECTRICAL CHARACTERISTICS

Figure 3.14 Reverse Biased PN Junction

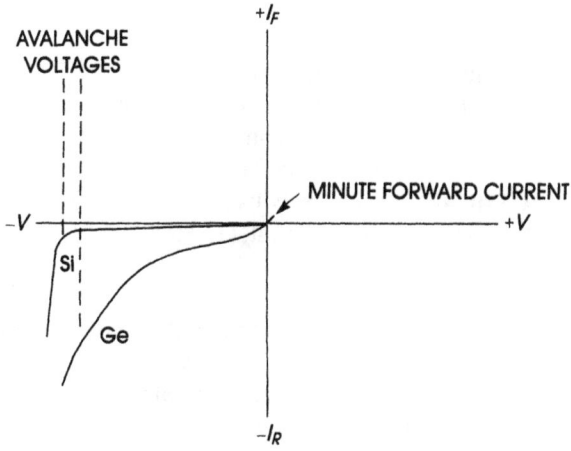

Figure 3.15 *I-V* Curve for Reverse Bias

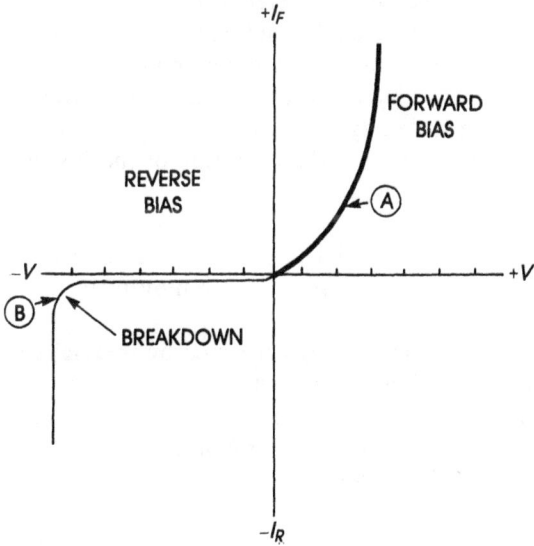

Figure 3.16 *I-V* Characteristics of a PN Junction

In this expression:

I = junction current in A
I_R = saturated value of reverse current in A
q = charge of the electron, 1.602×10^{-19}C
V = potential difference in V
T = temperature in kelvins (K)
k = Boltzmann's constant, $(1.380 \times 10^{-23}$ J/K)

This equation is valid for both positive and negative voltages and, thus, produces the curve shown in Figure 3.16. Typical values are 0.7 V for point A, which is the point where the forward-biased diode begins to draw noticeable forward current; and approximately 100 V for point B, where avalanche or zener breakdown occurs. In the forward-bias region of the device, it can be seen how the current rises rapidly for voltage increases once point A has been reached. Current in the reverse direction, however, is very low until sufficient reverse voltage is applied to cause breakdown. At this point (point B) there is excessive current drawn very rapidly. Under normal operating conditions you should avoid any excessive voltages in *either* the forward or reverse directions. With these voltages very high current and resulting high temperature caused by the increased current may, and probably will, permanently damage the device.

If we were to summarize the operation of the PN junction diode we have been discussing we could say:

- Joining *p*-type material to *n*-type material establishes a contact-potential hill which limits random movement of majority carriers across the junction.
- Minority carriers, usually generated by thermal energy, can easily cross a reverse-biased junction. Their flow is essentially independent of the magnitude of the bias, unless breakdown occurs.
- With forward biasing, the height of the potential hill is reduced and majority carriers cross the junction is greater numbers.

Before leaving discussions on the PN junction, two terms need to be presented and defined. Both terms are critical for microwave operation of solid-state devices. The terms are *junction capacitance* and *transit time*.

The first term, junction capacitance, has been mentioned frequently throughout our discussions of the PN junction. You will recall that we stated that junction capacitance has more meaning when the PN junction is either forward or reverse biased. It should now be clear what was meant by that statement, following the discussions presented on unbiased, forward, and reverse biased junctions.

Junction capacitance is, basically, what the name implies. It is the capacitance set up by placing a *p*-type material and an *n*-type material in contact with one another and either applying a bias voltage to them or letting the junction remain in a unbiased state. You will recall from the basic ac theory that a capacitor was produced when two conductive plates were separated by a dielectric. The spacing of the plates (thickness of the dielectric) determined the amount of capacitance. A relationship for this definition is

$$C = 0.0885 \ \epsilon_r \ \frac{(N-1) \ A}{t}$$

where

ϵ_r = the dielectric constant of the dielectric
N = number of plates
A = area of one side of one plate in cm
t = dielectric thickness (plate spacing) in cm

You will recall from our discussions of unbiased, forward, and reverse-biased PN junctions that the junction capacity varied exactly as predicted by the relationship alone. That is, an unbiased junction had a value of capacitance determined by the material joined and the resulting space-charge region; the forward-biased junction had a much smaller space-charge region (t decreased) and also a resulting higher junction capacitance; and the reverse-biased junction had a very wide space-charge region (t increased) and a correspondingly lower value of junction capacitance. Thus, the previous theory has been substantiated. There are two statements that can be made on junction capacitance:

1. It is present across *every* PN junction,
2. It is caused by the uncovered charge layers that come about as a result of the ionization of impurities.

For microwave applications the junction capacitance must be very low. Large values of junction capacitance will limit the high frequency operations of devices or may cause positive feedback within a device that will result in oscillations. When choosing devices for microwave operation, therefore, be sure that the junction capacitance of the device is minimal or very small.

The final term to be investigated is the one which made solid state devices much more desirable than vacuum tubes — *transit time*. In a semiconductor this is the time it takes for carriers to cross the depletion or space-charge region. It is a *delay* (τ) which is encountered as a carrier travels from one material to another. The effect of transit time is identical to that in vacuum tubes, although its operation is different, since one has a vacuum medium and the other has a dielectric medium. In traveling across a solid-state junction, the holes or electrons drift across with velocities determined by the resulting ion mobility, bias voltages, and junction construction. When more than one junction is concerned, as in a transistor, there are a combination of conditions which add up to form the total transit time. This total is

$$\tau_{ec} = \tau_e + \tau_b + \tau_{si} + \tau_c$$

where:

τ_{ec} = emitter-to-collector delay
τ_e = emitter-barrier changing time (delay)
τ_b = base transit time
τ_{si} = collector transit time
τ_c = collector charging time

- The emitter barrier charging time is an RC product of $r_e C_e$, where r_e is the emitter series resistance (determined by temperature, charge of electrons, and emitter current) and C_e is the emitter barrier capacitance.
- The base transit time is an expression determined by W_b (base layer thickness), D (minority carrier diffusion coefficient), and n (a factor related to the drift field in the base region associated with the amount of impurities).
- Collector transit time is determined by the time of carriers crossing the depletion layer, the drift velocity of the carriers, and the total width of the depletion layer in the collector region.
- The collector charging time is once again an RC product. This time it consists of r_c, the collector resistance and C_i which is the collector depletion-layer capacitance.

All of these parameters combine to produce an overall transit time (delay) through the device. Further discussions of two-junction, three-element (base-emitter-collector) devices will be presented in Chapter 5 on microwave transistors.

Thus, the PN junction has been formed and characterized. It can be seen that when two types of materials (*p* type and *n* type) are joined, a junction is formed which exhibits electrical characteristics which can be modified by application of bias voltages. Such parameters as junction capacitance and transit time are parameters which are critical when operating solid-state devices (with PN junctions) at microwave frequencies. Thus, when choosing the device for your particular application, be sure to consider these and other parameters which will aid the PN junction in doing the best job for your microwave application.

3.2 SCHOTTKY BARRIER JUNCTION

If you were to look up the definition of a Schottky barrier in an electronics dictionary, you would find words which say:

A simple metal-to-semiconductor interface that exhibits a nonlinear impedance.

A large portion of that definition is absolutely true. One part is in question. That part is the word *simple*. The Schottky barrier junction is not simple in terms of operation, it is only simple in the fact that a metal and a semiconductor are joined together and produce one of the most useful junctions ever devised for microwave applications.

As stated above, the Schottky barrier junction is set up by contact between a metal and a semiconductor. It gets its name from early analysis in an area termed *majority-carrier rectification* by W. Schottky in 1938. The semiconductor used is generally *n*-type silicon or *n*-type GaAs. This metal-semiconductor junction has many advantages over a PN junction when used at microwave frequencies. When the junction is forward biased, current flows because of majority carriers moving from the semiconductor to the metal. Minority carrier effects are virtually eliminated with this type of junction. This means that there is no reverse recovery time in a Schottky junction (since only majority carriers are used) and, then, there is no charge-storage capacitance. Low capacitance, as we have discussed previously, is an excellent property for microwave devices, since it allows them to operate at higher frequencies. The reliance on only majority carriers to perform the necessary operations within the junction classifies the Schottky, and any devices using it, as *unipolar*. This is as opposed to the *bipolar* effects of the PN junction.

Let us now take a look at the energy levels of a Schottky barrier junction and find out more about its operation. Figure 3.17 shows energy diagrams for metal (a) and a semiconductor with no surface charge (b). In the case

VACUUM

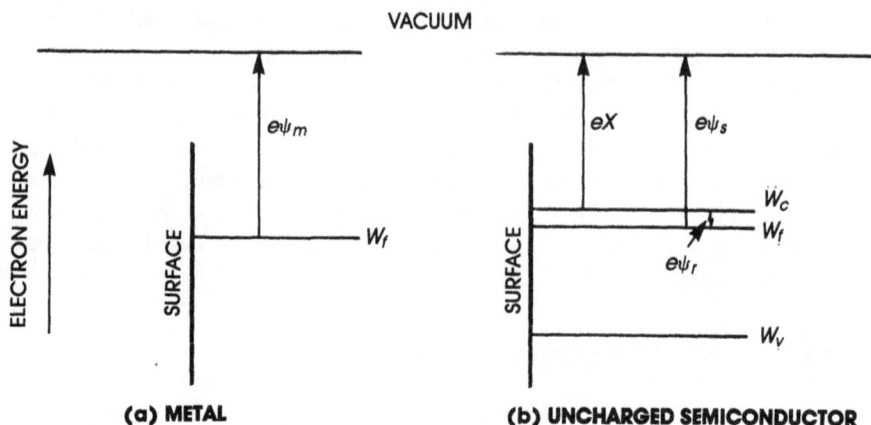

Figure 3.17 Energy Diagram for Metal and Uncharged Semiconductor

of the metal, ψ_m is the work function of the metal. This is a unit of negative charge for this discussion. You may recall this designation from our discussion of the PN junction. At the time ψ was used to designate a unit-positive charge. Our usage here, however, is negative and denotes the energy required to remove an electron from the Fermi level (W_f) of the metal and place it at rest in free space (vacuum). The factor e is used to convert the charge ψ_m to energy units. This work function is a constant value for a given atomically clean metal but varies as surface contamination increases. This is the only factor which governs its value. This is as opposed to the work functions of a semiconductor, ψ_s, which not only varies with surface contamination but also depends on the positions of the Fermi level and upon carrier density. These factors can be seen in Figure 3.17(b). (W_c is the wave function in the conductance band and W_v is the work function in the valence band.) One parameter which is not affected by semiconductor doping (or contamination) is termed X, the electron affinity. This is the amount of energy released or absorbed when an electron is added to a neutral atom. Its unit of measure is electronvolts and will increase with smaller (decreasing) atoms. This indicates that the electron affinity is greater for higher group numbers (Groups V, VI, VII).

The semiconductor depicted in Figure 3.17(b) is one without a surface charge, that is, a neutral state of the semiconductor exists. This, however, is a condition with is *not* a common occurence, and, when it does occur, it does not last for any period of time. What is a common occurence in a semiconductor is shown in Figure 3.18. This is a semiconductor with a net negative charge. In this state, positively ionized donors are formed just below the surface and distort the band edges are shown. Since the material is still at equilibrium, there is no net current and the Fermi level is still constant throughout. The distance of the Fermi level for the conduction load is designated as the amount of bending of the band edges and is shown as ψ'_{ns}. These are shown in Figure 3.18.

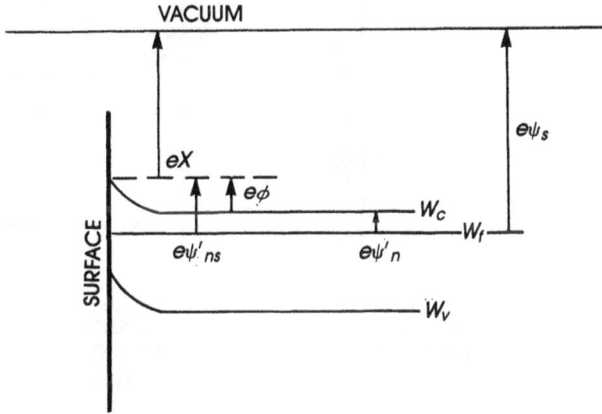

Figure 3.18 Semiconductor with Net Negative Charge

With the basic concepts of both the metal and the semiconductor presented, it is now time to place the two in contact with one another and form the junction. Figure 3.19 shows the two individual materials (metal and semiconductor) prior to their attachment to one another. It is important to notice the individual energy levels and work function placements before contact. Notice that the Fermi levels are not coincident in both materials. This is because, at this point, the two materials are not in equilibrium with one another. The figure also shows that there is some net surface charge present in the semiconductor, since there is band bending present.

Figure 3.20 shows the energy levels of the Schottky barrier as the metal and semiconductor are brought into contact with one another. Equilibrium has not been established, and the Fermi levels are not in line. A change in potential designated $\psi_m - \psi_s$ occurs across the gap. This is termed Δ in the figure. This charge is called the *contact potential*. As the surfaces are brought closer together, the field between the two surfaces, designated by $\psi_m - \psi_s / \delta$ will increase. The gap distance δ is in the order of 0.5 to 5 Å, depending on surface preparation and method of contact. This is a distance of 0.5×10^{-8} cm (1.96×10^{-9} in) to 5×10^{-8} cm (19.68×10^{-9} in). This thin layer is virtually transparent to carriers coming between the metal and the semiconductor. At its maximum height, the barrier which is seen by the carrier

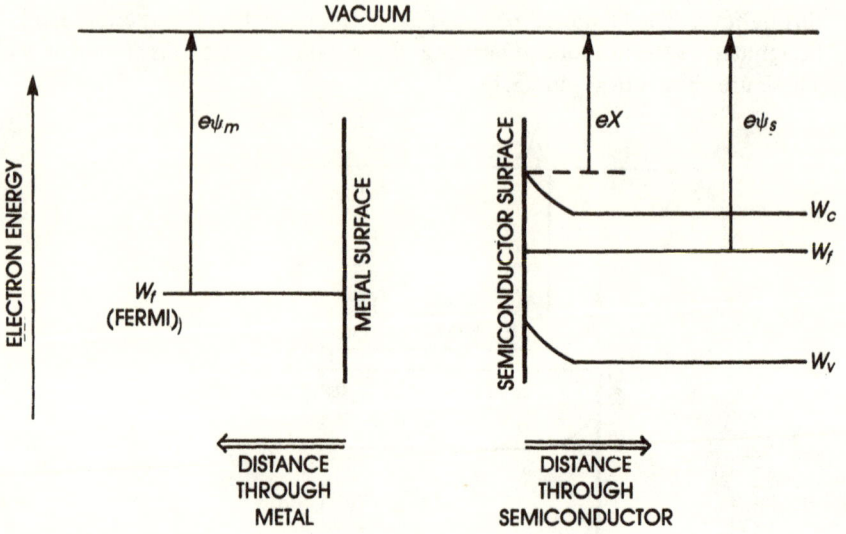

Figure 3.19 Metal and Semiconductor Before Contact

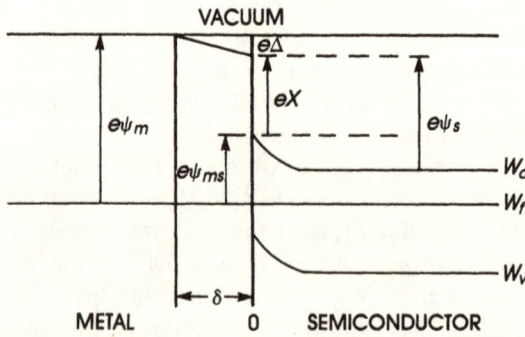

Figure 3.20 Metal and Semiconductor in Contact

is known as the *metal-semiconductor barrier height* and is designated as:

$$\psi_{ms} = \psi_m - X - \Delta$$

where

ψ_{ms} = work function of the metal-semiconductor
ψ_m = work function of the metal
X = electron affinity
Δ = change in potential across the gap

Table 3.1 below shows values of ψ_{ms} for various metals when contacted with Si (Silicon) and GaAs (Gallium Arsenide), both N-type semiconductors. (Values are in volts.)

Table 3.1

Metal	Si	GaAs
Au (Gold)	0.81	0.90
Ag (Silver)	0.69	0.88
Al (Aluminum)	0.67	0.80
Cu (Copper)	0.71	0.82
Pt (Platinum)	0.85	0.86
Ni (Nickel)	0.66	—
W (Tungsten)	0.69	0.80

Notice the heights of the barriers using silicon as opposed to those in gallium arsenide. All values are lower for silicon, which is why gallium arsenide is used for transistor application when the Schottky junction is advantageous for high frequency operation. Generally, diodes will use silicon for the Schottky junctions, since barrier heights need not be as high for diode operation. Diodes are covered in Chapter 4, while transistors (GaAs FETs) are covered in Chapter 5 of this text.

The maximum width of the barrier is that of the depletion layer. This is usually on the order of thousands of angstroms.

With the metal-semiconductor Schottky barrier junction analyzed and formed, it now remains to place bias voltages on the junction to produce the desired operation. Figure 3.21 shows the three conditions possible for a Schottky junction; no bias, forward bias, and reverse bias. Figure 3.21(a) is the familiar energy diagram we have been discussing throughout this session. It is the zero-bias condition and is reproduced in this figure to be used as a reference. It can be seen from the bias conditions presented that the bias voltage causes changes in the energy band bending and, more importantly, a change in the Fermi level (W_f) within the junction. In Figure 3.21(b), the forward bias condition, the bending is not as severe as the zero bias condition and the Fermi level is raised slightly above that of the metal. This allows the junction to conduct majority carriers more easily. (V_A is positive for an applied forward bias — the positive voltage being applied to the metal.)

In Figure 3.21(c) the reverse-bias condition is shown. In this case the energy band bending is extensive and the Fermi level of the semiconductor is much lower than that of the metal level. This basically eliminates any condition across the junction. (V_A is negative for this condition.)

Thus the Schottky junction has been shown for applications in the microwave area. Its operation as a *unipolar* device which relies only on majority carriers for conduction allows it to be used at much higher frequencies than its PN junction counterpart (a *bipolar* device).

The junction is used for diodes which are used in detection and mixers and has become the heart of GaAs field effect transistors and the gate junction. For this Schottky gate, certain metals such as aluminum, tungsten, platinum, and molybdenum will form the desired diode suitable for a gate junction when evaporated on GaAs. (All of these metals were described in Table 3.1, when we were discussing the work function, ψ_{ms}, of the metal-semiconductor combination.) When using any of the last three metals (molybdenum, platinum, tungsten) a layer of titanium is first put down to act as an adhesive, since they will not stick to GaAs on their own. A layer of gold gives the gate area the needed conductivity. Thus aluminum gates are much simpler to fabricate than other metals, even though other materials, such as titanium-tungsten-gold (TiW-Au) gates have advantages which are:

1. *Not* subject to *purple plague*, a brittle alloy which is formed when an aluminum-gold (Al-Au) bond is exposed to extended heat — as in the gate formation process.
2. Less subject to failure from electric discharge.
3. Can be plated an arbitrary thickness to reduce resistance (and reduce the device noise figure).

Thus it can be seen that, as in all devices, trade-offs are necessary when utilizing the Schottky barrier junction.

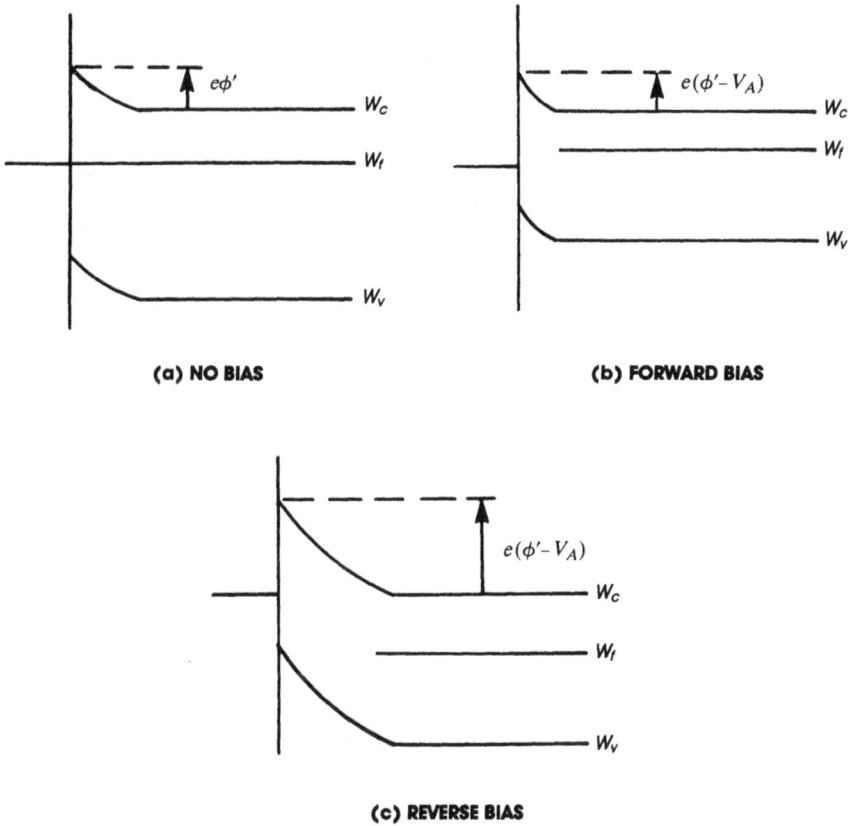

(a) NO BIAS

(b) FORWARD BIAS

(c) REVERSE BIAS

Figure 3.21 Schottky Barrier Bias Condition

3.3 SUMMARY

This chapter has covered the two types of junctions utilized in microwave solid-state devices — PN junctions and Schottky barriers. Both types have applications throughout microwaves with the Schottky finding many new applications as the technologists advance and gallium-arsenide (GaAs) devices become more prominent. The Schottky lends itself much more easily to GaAs than does the PN junction. Thus, the area of solid-state devices which actually causes them to operate and perform at microwaves has been presented — the solid-state junctions.

Chapter 4
Microwave Diodes

When most people think of a diode they picture a small glass or plastic-packaged device with two leads on it used for rectifying ac in a power supply or clamping a pulse to a specific dc level. This representation, of course, is very accurate for many of the low frequency, dc, and logic applications encountered throughout electronics. When we speak of *microwave diodes*, however, we are speaking of a two-element device which will do much more than rectify or clamp a dc voltage.

To understand the complexity and versatility of microwave diodes, consider what functions they can perform. This two-element service is able to switch, attenuate, mix frequencies, detect, amplify, oscillate, and modulate signals in both the microwave and millimeter bands of frequencies. This small device which is defined by the dictionary of electronics as *a two-terminal electronic device that will conduct electricity much more easily in one direction than in the other*, can perform functions which were unheard of with three-terminal, or any other, devices as little as thirty years ago. Obviously, we are talking about devices which have characteristics that involve much more than "conducting electricity much more easily in one direction than in another." We are speaking of devices which use silicon or gallium arsenide as a semiconductor; which may use a PN junction or Schottky junction for operation; may have an intrinsic layer added during construction to control the resistances of the device; will exhibit negative characteristics; or may not even rely on semiconductor junction properties for their operation. These are the characteristics of a variety of microwave (and millimeter) diodes. Not, as you can see, the typical characteristics used in the basic definition of a diode. This is because a microwave diode is much more than just a two-element device which has limited capabilities. It is a complex device which is an integral part of many sophisticated microwave circuits and systems and, as such, deserves close inspection and study to determine how these tiny marvels can be used to their greatest potential.

We will investigate six types of microwave diodes in this chapter. They are: Schottky, PIN, Varactor, Gunn, IMPATT, and TRAPATT diodes.

Each of these devices will be presented with an analysis of the device, theory of operation, construction, and applications. Where applicable, a typical data sheet will also be presented to illustrate common parameters, how to interpret them, and where (and how) they can be useful for your applications.

4.1 SCHOTTKY DIODES

You will recall from Chapter 3 that our discussions of the Schottky junction stated that, in contrast to the PN junction, the Schottky is a metal-to-semiconductor contact which makes it very useful at high frequencies due to its characteristic of using only majority carriers for current. This effectively eliminates any minority carrier effects which allows the device to operate in such circuits as microwave detectors or mixers with great efficiency, whereas the PN junction would have frequency limits due to its reliance on both majority and minority carriers for operation. The nonlinear impedance exhibited by the Schottky junction also is very useful for mixer operations, since a mixer depends on the diode being in a nonlinear mode to generate the sum, differences, and products of input frequencies.

As stated above, the properties of the operating Schottky diode are determined by the majority carriers only (electrons). This results in a device which can be switched rapidly from forward to reverse bias since only majority carriers are involved. There are no minority carriers to be concerned with, as with the PN junction. These minority carriers cause a time delay in the PN junction, since they are in a storage state rather than free to move around. This effect is not present in the Schottky junction, and, thus, there is no delay and switching is considerably faster.

In our discussions in Chapter 3 it was also stated that there is an energy barrier at the metal-to-semiconductor junction (called the *Schottky barrier*). This barrier exists because of the difference in the work function of the two materials. This barrier is unaffected by a reverse bias, but decreases substantially with forward bias. These conditions were shown in Figure 3.21, when energy levels of the Schottky barrier were analyzed. With this arrangement, a forward-biased Schottky diode will have the majority carriers (electrons) easily injected from the semiconductor material into the metal, where the energy level is now much higher. Once the electrons are in the metal, they give up their excess energy in a very short period of time. This time ranges in the order of 10^{-15} seconds or one femtosecond. To illustrate how fast this is, consider the fact that one femtosecond is one one-thousandth of a picosecond. When the electrons give up their energy in this short period of time, they then become a part of the free electrons of the metal. As previously stated, when the junction is reverse biased the energy level of the barrier is too high for the electron to overcome, and, thus, the device will not conduct. With this brief review of how the Schottky junction operates as opposed to the PN junction, let us now further investigate the Schottky diode, its theory, construction, and applications.

An equivalent circuit of a Schottky diode is shown in Figure 4.1. The figure shows two series components L_s and R_s. The term L_s is the series inductance present in the wire bonded to the metallized area placed over the junction to make contact with the diode itself. The term R_s actually is two terms R_{sc} (resistance of an epilayer) and R_{ss} (resistance of the substrate). These terms will be further elaborated on shortly. The Schottky junction as shown in the figure consists of R_j (junction resistance) and C_j (junction capacitance). The junction resistance is shown as a variable resistance, since operating conditions can cause this parameter to vary. This is why most diodes are characterized as having a specific R_j at a specific operating voltage. In low-level detector applications, for example, the R_j value is approximately equal to the small-signal, low-frequency diode parameters at usual detector levels. These are values given on a diode data sheet and can, for all practical purposes, be used for most microwave applications. The junction capacitance, C_j, is in shunt with the junction resistance, R_j, discussed above. This capacitance exists at the barrier of the metal-semiconductor combination. It is a term (parameter) which will also vary with applied bias voltage. The value can be obtained by

$$C_j = \frac{C_{jo}}{1 + \left(\dfrac{V}{\phi} \right)^\gamma}$$

where

C_{jo} = junction capacitance at zero bias
V = applied forward bias
ϕ = barrier voltage
γ = the exponent of the C-V relationship (C_j is typically from 0.3 to 0.5 pF).

The series resistance, R_s, as previously mentioned, is actually a combination of two resistances. These are R_{sc}, resistance of the epilayer, and R_{ss}, resistance of the substrate. This term is probably the most important of any resistance term when considering a diode for detector or mixer applications. If R_s is high, power can be dissipated within the diode, and, thus, a maximum transfer of power within the device cannot take place. When choosing a diode for those applications, one should choose a device which has a minimum value of R_s. The value of R_s (usually 4 to 6 ohms) will usually vary with

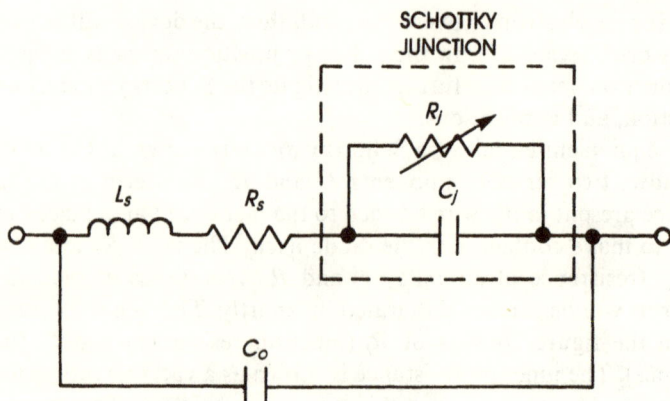

Figure 4.1 Schottky Diode Equivalent Circuit

bias voltage. For diodes with lower values of resistance this variation may be considerable. The variation within the device is a function of the concentration profile of the semiconductor material near the edge of the space-charge boundary and is related to the change in C_j with bias, since both quantities increase as a result of a decrease in space-charge width, as can be concluded from discussions in Chapter 3.

When a Schottky-barrier diode is forward biased the series resistance can be calculated as follows:

$$R_s = R_{so} + \frac{\epsilon \rho}{C_{jo}} \left[1 - \left(1 - \frac{v}{\phi} \right)^\gamma \right]$$

where

R_{so} = series resistance at zero bias
ϵ = permittivity of the semiconductor
ρ = average resistivity of the semiconductor epitaxial layer
C_{jo} = junction capacitances at zero bias
ϕ = barrier voltage
v = forward bias
γ = the exponent of the *C-V* relationship

So it should now be evident how important R_s is to the Schottky diode operation and how many parameters are involved when determining its value.

The final parameter to be discussed from Figure 4.2 is C_o. This is an overlay capacitance which is across the oxide layer of the device. The value

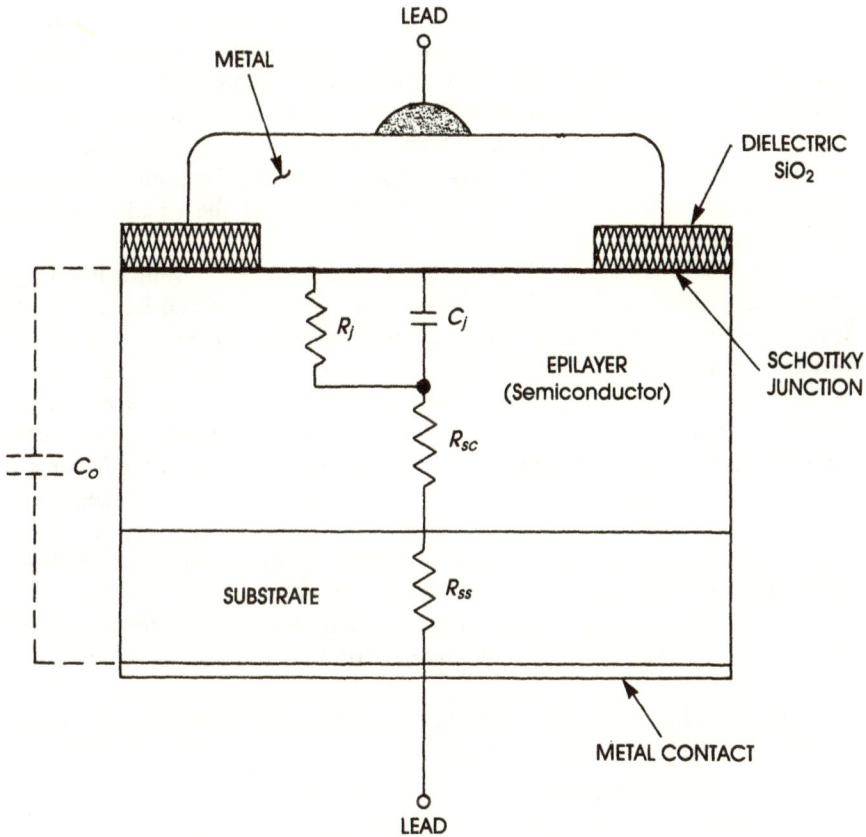

Figure 4.2 Schottky Diode Circuit

of this capacitance obviously depends on the physical size of the oxide layer as well as on its composition and characteristics. Generally this layer is either silicon (Si) or gallium arsenide (GaAs) with very predictable values.

The total capacitance, C_T, which should be considered for a Schottky diode, consists of the junction capacitance, C_j; the overlay capacitance, C_o; and the package capacitance, C_p. The final parameter, C_p, will depend on the style of package used and the methods used to attach the diode chip to the package. Also, if chip devices are used, this parameter becomes zero, since there is no package for the device.

As in any semiconductor device, the method of contact from the chip to the outside world is a critical issue. This contact can many times be the difference between a device being right for your application and having to look for another one. Basically, there are three ways used to make contact to a Schottky diode chip. They are *point contact (Whisker), C-spring contact,* and *thermocompression bond.* These are shown in Figure 4.3. You can see that the point contact connection shown in Figure 4.3(a) is just what the name implies — a very small point which is brought in contact with the metal of the Schottky junction (the black portion of all the diagrams in Figure 4.3). This is a very low capacitance bond because of the small area that actually makes contact.

The C-spring contact shown in Figure 4.3(b) is a very simple bonding technique which has a metallic strip bent into a C-shape and bonded to a lead at the metal contact of the junction. This makes a very economical bonding method which also has certain mechanical advantages as a bond which will flex without placing strain on the actual bonded areas. One disadvantage of this type of contact is the added value of cpacitance because of the longer contact and larger bonding area. For this reason the C-spring contact type diode is only used for lower microwave frequency applications.

The final contact method, the thermocompression bond, is shown in Figure 4.3(c). As can be seen, this is by far the most rugged of the bonding methods used. The bonding wire is "forced," by heat and pressure, into the metal area of the Schottky junction and becomes an integral part of it. This is the most widely used type of contact for the majority of microwave Schottky diodes.

In summary of the different type of contact methods we could categorize them generally as:

Point Contact (Whisker) — Used in millimeter applications, since the construction minimizes the parasitic capacitance.
C-Spring — This contact is used in most glass-package diodes. It is the least costly to manufacture.
Thermocompression Bond — Mechanically, the best type bond to use. Reliability is excellent.

With the Schottky chip explained and contact made to it, we now need to put the diode into a package. As we have implied before, the package type is many times as important as the actual Schottky chip itself. If the package is going to degrade the parameters you worked so hard to get within the chip, then you have chosen the wrong type of package. The right package is the one which will have a minimum effect on the chip's performance and could possibly enhance it. Figure 4.4 shows two typical packages used for Schottky barrier diodes. Figure 4.4(a) is a glass package which can be considered to be a general purpose case used for lower microwave frequency applications. This may seem like a contradiction from statements made previously. You will recall that the whisker type contact was characterized

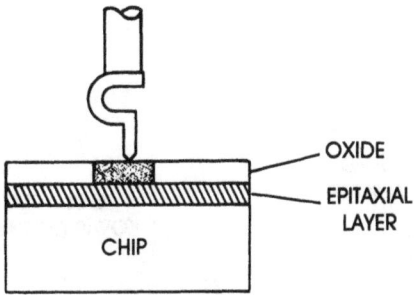

(a) POINT CONTACT

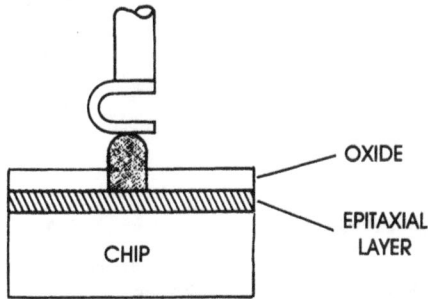

(b) C-SPRING

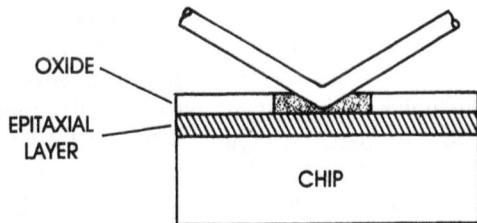

(c) THERMOCOMPRESSION BOND

Figure 4.3 Contact Methods

as being useful up into the millimeter range because of the low parasitic capacitance produced. Now we have a diode which uses the whisker contact and we say it is used for lower microwave frequency applications. To clarify this we can say that the whisker contact is excellent for millimeter usage,

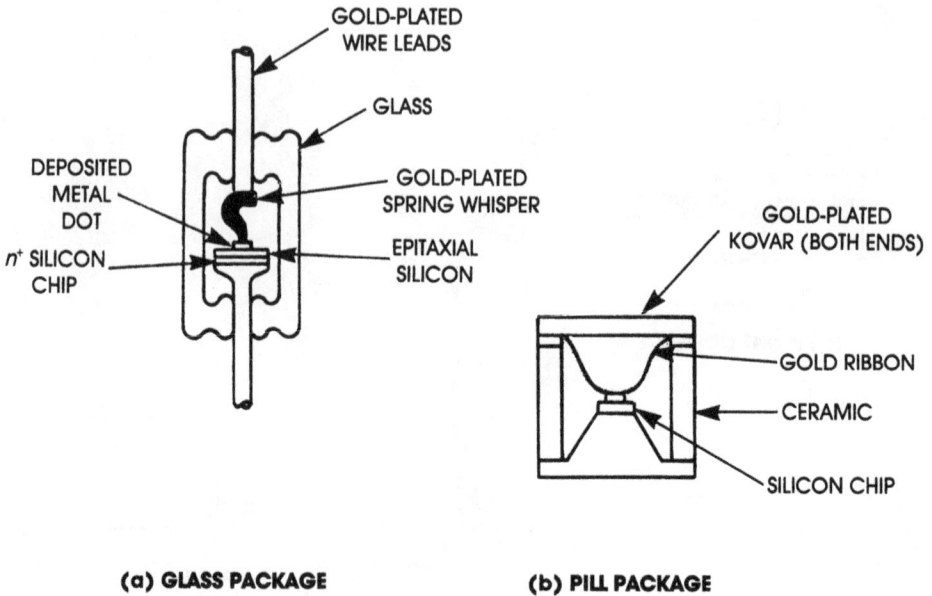

(a) GLASS PACKAGE   (b) PILL PACKAGE

Figure 4.4 Schottky Diode Packages

but when placed in a glass package with radial leads it loses all of the advantages that the whisker contact exhibited. Thus we have a case, as stated in our opening statements on packages, where a case style will degrade the parameters of the chip. So the glass package is a low microwave frequency device. When this package is used, the leads should be kept as short as possible. In this way the range of operation can be raised a certain degree.

Figure 4.4(b) is the pill package. This is a package which is used for many higher frequency microwave applications. You should not make a size comparison between Figures 4.4(a) and 4.4(b). The glass package is in the order of 0.200 inch to 0.250 inch, whereas the pill package is in the order of 0.050 inch to 0.070 inch square. Thus the pill package is excellent for miniature circuits which are operated in the higher microwave range. Notice also that the device uses the very highly reliable thermocompression bond. This ensures that, even though the diode package is small and compact, it is also very rugged.

Figure 4.5 shows three other packages which are available for Schottky diodes. Figure 4.5(a) is termed a *coaxial* package. This is because it fits into a coaxial line, rather than being soldered or bonded to a piece of stripline or microstrip. Figure 4.5(b) is a miniature package which finds most of its applications where stripline and microwave circuits are involved at higher

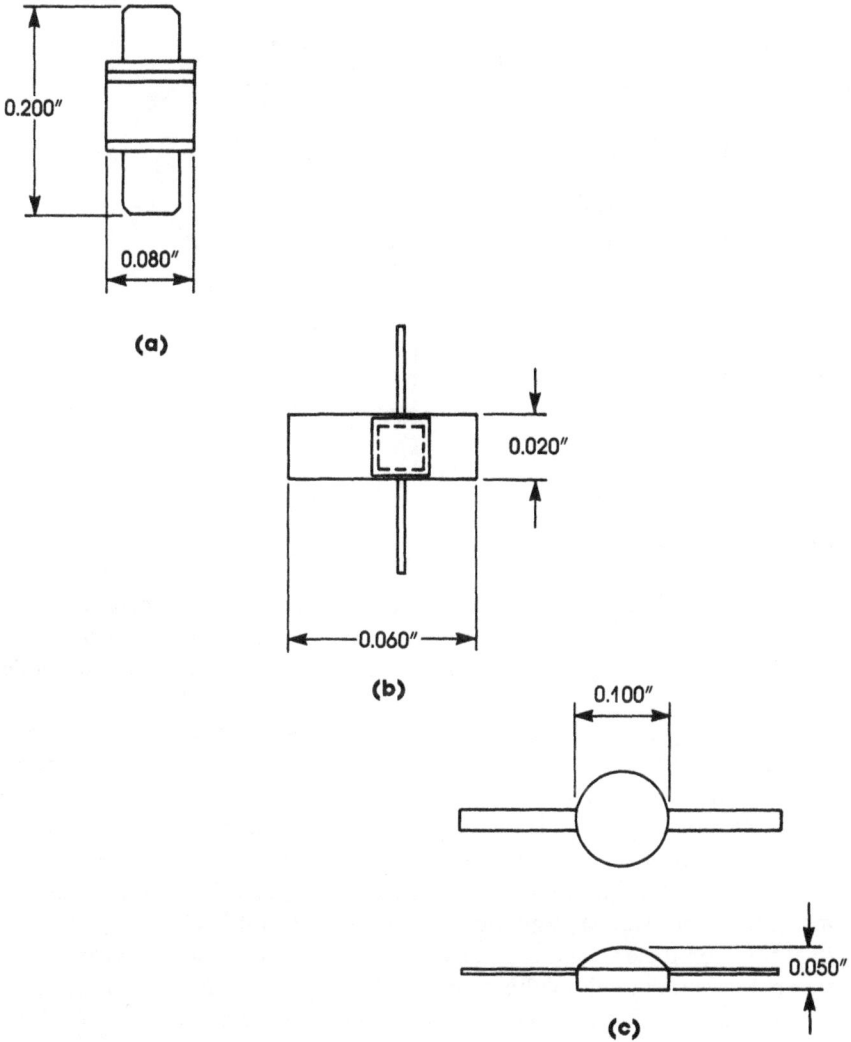

Figure 4.5 Schottky Diode Packages

microwave frequencies and into the millimeter range. Figure 4.5(c) is probably the most widely used and most recognizable package used for Schottky diodes. It is a general purpose package which is used for microwave applications up to 10 GHz.

Figure 4.6 is a representation of a *quad-ring* configuration. This is a package where four Schottky diodes are put in one package and connected

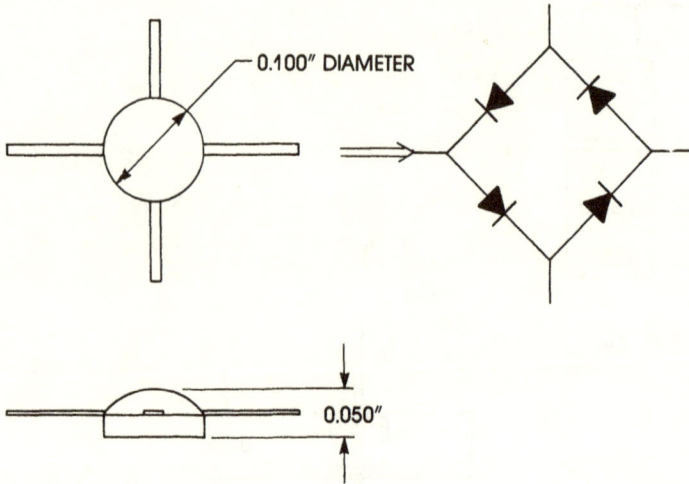

Figure 4.6 Quad Ring

as shown. This is an excellent package for use in microwave mixer circuits. All of the required diodes are in one package which conserves space and provides much better parameters as opposed to using four separate diode packages.

Some of the applications of Schottky diodes have been mentioned throughout our discussions. We will now concentrate on the two primary applications — mixers and detectors.

The Schottky diode is the very heart of the microwave mixer. As can be seen in Figure 4.7, the diode assembly, as we refer to it, receives signals from both the RF input and the local oscillator (LO). The diode assembly may be anything from a single diode in each leg to a multidiode configuration, depending on the application. The diodes are driven into a nonlinear region by a combination of bias voltage and local-oscillator input level. Many times only the local-oscillator level is relied on to produce the nonlinearities. This nonlinear operation is an absolute requirement for the mixing process to occur. To understand this further, consider what would happen if you took an ordinary microwave amplifier and applied two signals at its input, one at 4 GHz and one at 4.1 GHz, for example. If the amplifier was one which passed more than a single frequency, which most are, the signals would be passed through and amplified according to the transfer function of the amplifier at each particular frequency. This would be true if both signals were of a sufficiently low level not to saturate the solid-state devices within the amplifier. The result would be two signals displayed on the spectrum analyzer, one at 4.0 GHz and one at 4.1 GHz, at their appropriate levels.

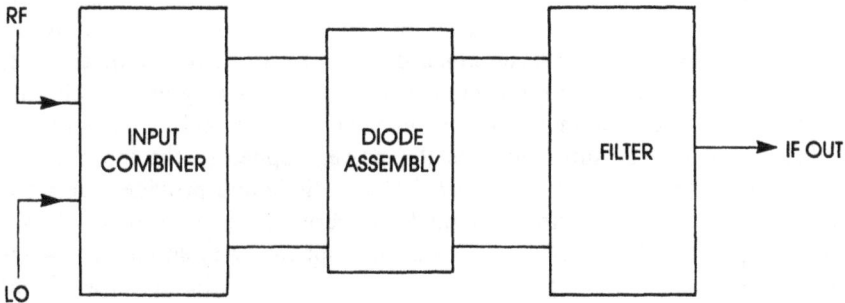

Figure 4.7 Mixer Block Diagram

Suppose now that we kept the 4.0 GHz signal at the previous level and raised the 4.1 GHz signal level to something on the order of +13 dB (20 MW) as opposed to its previous level of -25 dB, for example. The result would be that the 4.0 GHz signal still would allow the amplifier to operate normally, but the 4.1 GHz signal would be overdriving and saturating the amplifier. This would cause the solid-state device to be driven out of a nice comfortable linear region where a certain input level gives an orderly and predictable output level, and into a nonlinear region where a group of unknowns can and do exist. What is now present on the spectrum analyzer is the original group of signals of 4.0 GHz (f_1) and 4.1 GHz (f_2) plus one at 8.1 GHz ($f_1 + f_2$), one at 0.1 GHz ($f_2 - f_1$), one at 12.1 GHz ($2f_1 + f_2$) and a barrage of other combinations of f_1 and f_2. We have created a device which mixes the two input signals, and it only remains to choose which signal output we desire. This is done by means of the filter shown in Figure 4.7. Most of the time the lower frequency signal is desired (0.1 GHz in the case shown alone), so a low-pass filter does a very nice job of getting rid of the rest of the unwanted harmonic and mixing combinations.

So you can see how important the choice of a Schottky diode can be for proper operation of a mixer circuit. It must be one which can be made to mix (driven into nonlinear operation) easily and still have low-noise characteristics itself, so as not to degrade the overall mixer performance once mixing has been accomplished. Also, the junction characteristics of the diode R_j and C_j, must be of such a level so as not to cause excessive conversion losses throughout the mixer. (Conversion loss is the loss through the mixer from the input RF level to the output IF level.) As was previously stated, the R_j in conjunction with R_s (series resistance) can control the characteristics of the Schottky diode. Typical values of C_j can run from 0.3 pF to 0.8 pF. Values of R_s are in the range of from 4 to 6 ohms for some diodes with values up to 25 ohms allowed for some detector applications.

The second application, the detector, is one of the simplest of circuits to visualize, but one of the most useful and relied upon in the microwave field. Figure 4.8 shows a basic block diagram of a microwave detector. It consists of some form of input matching circuit. This may be an elaborate two or three section matching arrangement or a simple *L-C* network. Whatever it is, it ensures that the RF power applied to the input of the detector will be efficiently transferred to the diode and provide a detected output. One statement should be made concerning this input block. It is a matching circuit for specific conditions and not for every situation imaginable, that is, the matching network is designed for specific power level and output impedance levels. The reason for this is that scattering parameters (S-parameters) are used to characterize the diode. These parameters are valid only at certain power levels. If the level is changed, more than a little, the S-parameters change, and, thus, the matching network. Therefore, when a *detector assembly* is characterized, you will see a frequency range, power range, and output impedances (or load impedances) at which the data were taken. So be aware that the detector diode is characterized under specific conditions.

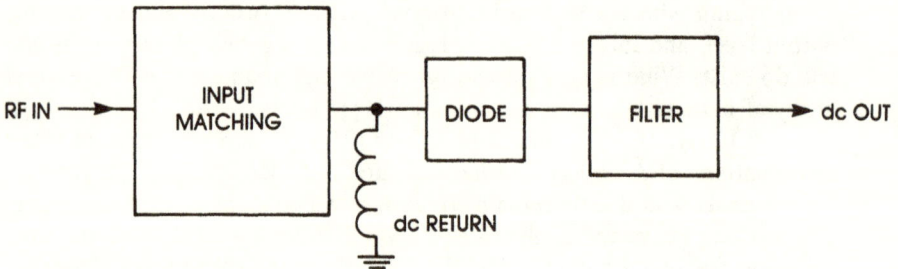

Figure 4.8 Detector Block Diagram

The detector finds many applications in microwaves. Figure 4.9 shows three of them. The first is a monitoring system in which a portion of a microwave signal is sampled by a directional coupler, detected by the detector, and displayed on a meter. This is a very common application of a detector. Figure 4.9(b) is an application used when a microwave power meter or a spectrum analyzer is not available. In this case you can place the detector at the output of a device being tested (being careful not to have too much power at the input to the detector) and see a deflection on the scope which is proportional to the circuit output. Figure 4.9(c) shows an automatic leveling circuit (ALC) loop. This is used to ensure that a leveled

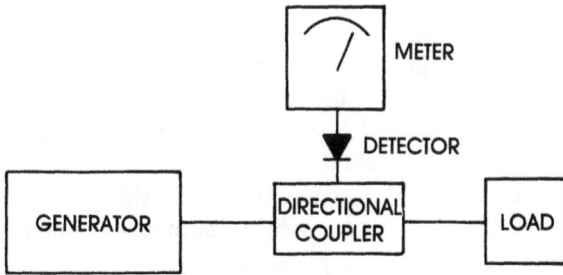

(a) MONITOR CIRCUIT

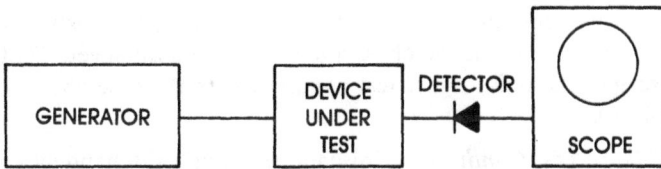

(b) USE WITH AN OSCILLOSCOPE

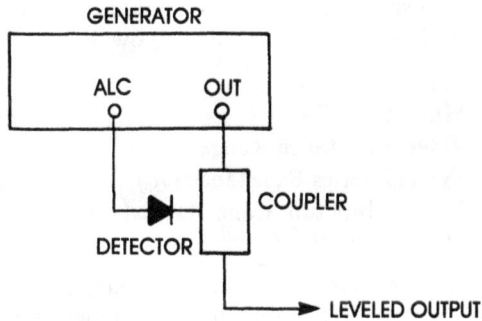

(c) AUTOMATIC LEVELING LOOP

Figure 4.9 Detector Applications

power is available over a swept range of frequencies. Any variations (up or down) coming from the generator are sensed by the coupler-detector combination. The resulting voltage change is sent back to the ALC circuitry (usually a PIN attenuator — to be discussed in the next section), and the output of the generator is adjusted accordingly. Thus, the output level of the generator can be held very constant.

With applications discussed, it would do well at this point to present a typical data sheet for Schottky diodes used for both mixers and detectors. The sheet shown below is for mixer applications.

Test Frequency	9.375 GHz
Noise Figure	6.5 dB (max)
VSWR	1.5:1
IF Impedance	250-450 Ω
Series Resistance (R_s)	8 Ω (typically)
Junction Capacitance (C_j)	0.1 pF (at zero volts)

The terms presented have all been presented previously. The only explanation needed is one which accompanies the data sheet. That is one which describes the noise figure test conditions. In the case above it would read as follows:

Noise Figure Test Conditions: Noise figure is single sideband measured with 30 GHz IF; noise figure of IF = 1.5 dB max; LO power = 1 MW; excess noise ratio = 15.3 ± 0.5 dB @ 9.375 GHz.

This will allow someone else to duplicate all of the measurements presented in the data sheet.

A data sheet for a Schottky diode to be used in a detector circuit is somewhat different. One is shown below:

Test Frequency	10 GHz
Minimum TSS	-52 dBm
Video Resistance Range	1.2 - 1.8 KΩ
Typical Series Resistance (R_s)	25 Ω
Typical Junction Capacitance (C_j)	0.07 pF
Min Sensitivity	3500 mV/mW

As in the case of the mixer diode, some parameters need further explanation. The first is junction capacitance and would be:

• Capacities measured at 1 MHz and 0 V.

The second is minimum TSS (tangential signal sensitivity) and would read:

• Video bandwidth is 2 MHz; video amplifier noise R = 500 Ω; input impedance = 10 kΩ: dc bias = 20 μA.

These conditions, once again, would be necessary to duplicate the published results.

Before proceeding further, it would be beneficial to explain what TSS is in a detector circuit. As stated previously, TSS is *tangential signal sensitivity* and is shown in Figure 4.10. TSS is a direct measure of the signal to noise voltage as shown in the figure. The measurement is conducted by applying a pulse signal to the input of the detector and displaying the output on an oscilloscope (Figure 4.10 is the oscilloscope display). The level of the pulse (input power level) is decreased until the highest noise peaks with no signal are the same as the lowest noise peaks when a signal is present. That is, height *BC* is equal to the height *AB*. This condition corresponds to a signal to noise ratio of 2.5, or 8 dB (NF_{dB} =20 log ratio). The level of the input when this condition occurs is read as TSS (-52 dBm on our data sheet). What this is saying is that you can have an RF signal at -52 dBm at the input to your detector and still have a usable output voltage. Any input level lower than this will produce a voltage which is more system noise than usable voltage.

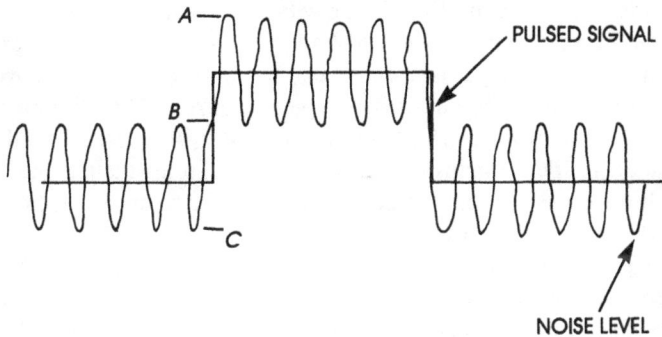

Figure 4.10 TSS Presentation

The final term that should be clarified is another *sensitivity* term in the data sheet. This one is *minimum* sensitivity. With this term an RF signal of known power level (mW) is applied to the input of a detector. The resulting voltage at the output is recorded (mV). This is repeated for various power levels and plotted until a curve of power input *versus* voltage output is obtained. Over the linear range of this curve it can be determined that you will get a certain voltage out for a certain amount of power. That is, for every milliwatt of power you can expect 3500 mV output voltage (in the case of our data sheet). This is valuable information when you are using the detector in conjunction with other circuits that may have critical parameters of either power or voltage.

Thus, the Schottky barrier diode is set down as the general purpose microwave diode. Its primary applications have been described as for mixers and detectors, although other applications are possible.

4.2 PIN DIODES

A PIN diode is defined as one which is made by diffusing the semi-conductor with p dopant from one side and n dopant from the opposite side with the process so controlled that a thin *intrinsic* region separates the p and n regions (thus, *PIN*). The storage time of the PIN diode is long enough that it cannot rectify at microwave frequencies. Instead, it behaves like a variable resistor with its resistance value controlled by a dc-bias current. This is why you will hear the PIN diode being referred to as a *microwave variable resistor*.

The key to the operation of a PIN diode is the intrinsic layer. An ideal intrinsic layer, or pure condition, is one which is not achievable. Such a layer would contain no doping and would have a resistance on the order of 300 kΩ-cm. In reality, the layer is not intrinsic in the true sense of the word but does in fact contain some impurities, typically boron. These impurities can be adjusted to the order of 10 kΩ-cm or generally 1000 Ω-cm. To illustrate this point of an ideal intrinsic layer as opposed to an actual one, we will consider both possibilities. It is customary to refer to the ideal layer as an *i* layer and the actual intrinsic layer as the π layer.

Figure 4.11 shows two cases referred to above. Figure 4.11(a) is the ideal case where the intrinsic layer is ideal. At zero bias, the diffusion of holes and electrons across the junction causes space-charge regions of thickness inversely proportional to the impurity concentration to form in the p and n layers adjacent to the i layer. For the ideal case, which is shown in Figure 4.11(a), the i layer has no impurities. Thus, the i layer is depleted of carriers. This results in a region of fixed negative charge in the p layer and a region of fixed positive charge in the n layer. There will be equal charges in the two layers and a minimal positive charge in the i layer as shown. Ideally, this charge will be zero. At this time the PN junction of the device is effectively at the p layer of the device as shown.

When you consider the practical case, or that which has the π layer, there is a different situation. With no bias the diffusion of holes and electrons across the n-π junction would produce a very thin depletion region in the n layer and a thicker depletion region in the π layer, containing equal but opposite fixed charges as shown in Figure 4.11(b). The PN junction at this

(a) The *p-i-n* TYPE

(b) The *p-π-n* TYPE

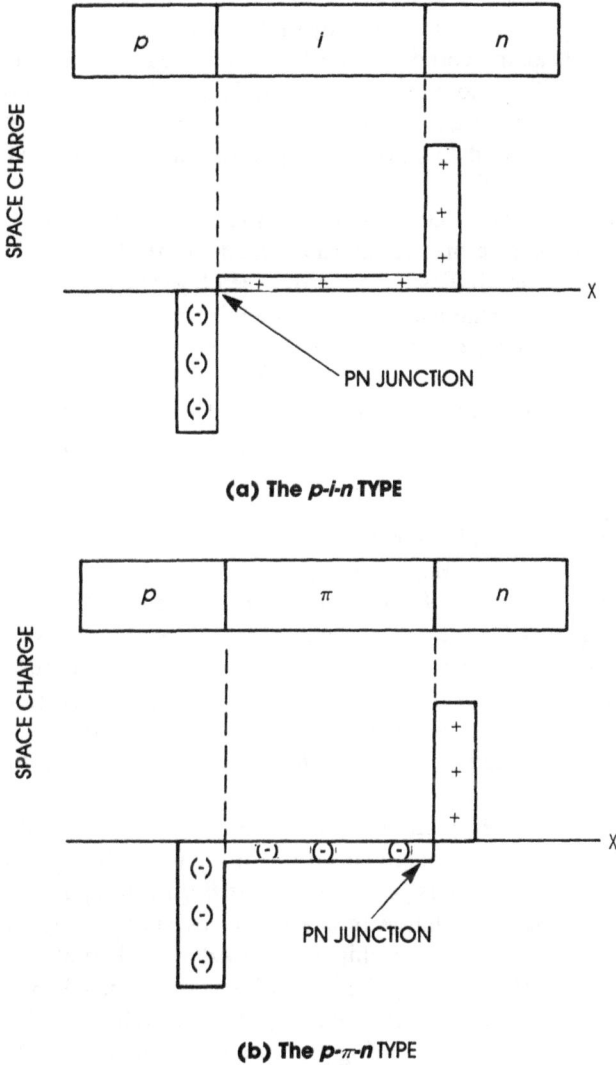

Figure 4.11 The *i* and *π* Layers

point is effectively at the *n* layer side of the device. As reverse bias is applied, the depletion regions become thicker until the entire *π* layer is swept free of mobile carriers. The applied bias necessary for this to occur is referred to as *the bias required to sweep out the π region*. This is an important bias to know, since many microwave applications require that the π region be swept out so that any current flow by mobile carriers in the resistive π region do not provide an increased signal loss. If there are no mobile carriers (*swept clean*), there will be no additional losses. Thus, when PIN diodes are used in microwave switching applications biased in the *off* condition, the bias is generally well beyond this swept-out voltage, so that losses are kept to a minimum.

So it can be said that the i layer of the ideal diode is swept out at zero bias or at any reverse-bias value. In contrast to the ideal case, the practical diode requires a small bias to sweep out the i layer (π layer).

The PIN diode has a more complex equivalent circuit than the Schottky diode did in the previous section. This can be seen by referring to Figure 4.12. The first components of this diagram to be considered are those which result from connecting the leads of the device to the wafer and those associated with the package. These are L_s, C_p, and C_f. The term L_s is the total series inductance, most of which is concentrated in the leads themselves. C_p is a representation of the stray capacitance shunted across the wafer by the package. This can be minimized by careful choice of package type. The term C_f is called the *fringing capacitance*. This is the stray capacitance shunted across the wafer from the leads. It can be seen that this is the case since (in Figure 4.12) the physical capacitor, C_f, is only across the elements that make up the diode and does not include L_s, the lead inductance.

It is now time to look at the parameters purely associated with the wafer, or chip. The depletion, or junction, region of the device exhibits both a capacitance, C_j, and a resistance, R_j, as shown in Figure 4.12. The value C_j is present due to charge storage at the boundaries of the depletion region. The resistance of the junction, R_j, is actually the reciprocal of the conductance caused by carriers generated within the region. A third parameter which has not been mentioned as yet is termed the *diffusion capacitance*, C_d. This represents the storage charge in the region as current flows through the depletion region. When the chip is forward biased, mobile carriers are injected into the i layer during the forward portion of the RF cycle with some of them being extracted on the reverse portion of the cycle. The storage of charge which takes place during this time is explained by the diffusion capacitance.

The final parameter, R_s, is the sum of the resistance of the p and n layers and any resistance associated with the contacts to these layers. This is analogous to the R_s explained for Schottky diodes.

The values of some of the parameters listed above are very difficult to obtain with great accuracy. It is possible to do this, however, and when it is acomplished one fact is very evident. The parameters are very dependent

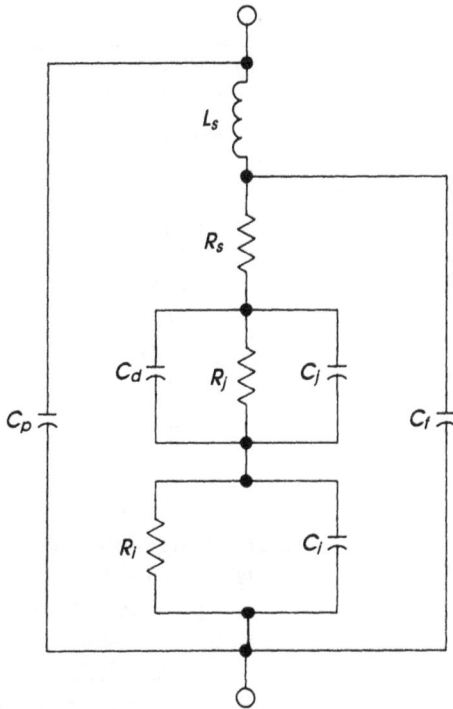

Figure 4.12 Diode Equivalent Circuit

on the method of biasing used. Table 4.1 illustrates just how much some of the parameters change with zero bias, 50 V reverse bias, and 25 mA forward bias.

It can be seen from the table above that such parameters as C_f, C_p, L_s, and R_s do not change with biasing. This is because, as previously stated, these parameters are the result of either the connection of leads to the chip or the resistance of the p or n layer and contacts to them. Obviously, bias has nothing to do with these values. The remaining parameters associated with the junction and the intrinsic layer are effected by bias. (The parameter C_d has no value placed in the 0 V and 50 V reverse columns because the value measured was negligible.)

With the basic theory of operation and the PIN diode parameters presented, we can now see how a PIN diode is constructed. We will present two basic types of construction (*planar* and *mesa*) and two variations of these

Table 4.1

Parameter	OV Bias	50 V (Reverse)	25 mA (Forward)
C_j	0.8 pF	0.17 pF	$\geqslant$10 pF
C_d	——	——	>3 pF
C_i	0.25 pF	∞	>0.25 pF
C_f	0.02 pF	0.02 pF	0.02 pF
C_p	0.3 pF	0.3 pF	0.3 pF
L_s	0.3 nH	0.3 nH	0.3 nH
R_s	0.3 Ω	0.3 Ω	0.3 Ω
R_j	>10^9 Ω	>10^9 Ω	0.1 Ω
R_i	2.5 KΩ	0 Ω	0.5 Ω

methods which individual manufacturers have fabricated to take advantage of the favorable characteristics of both the planar and mesa.

Figure 4.13(a) shows the *planar* construction used for a PIN diode. It is produced by diffusing boron through a window in the thermal oxidation passivation to form the junction. This type of construction does not find a lot of applications at microwave frequencies. Generally, this is the most common form of fabrication for low-frequency diodes.

For the planar construction to be considered for microwave applications we must be aware of two drawbacks which this form of construction has. First, the PN junction depletion region is shaped like a plane-parallel plate capacitor in the center of the device but is cylindrically shaped at the edges. Figure 4.12(b) illustrates this phenomenon very vividly. This cylindrical edge reduces the voltage breakdown of the device considerably. Second, the interactive silicon surrounding the junction produces extra fringing capacitance. This, as we have previously discussed, is not what we would like a device to do, if we are planning on using it in a microwave application. Excess fringing capacitance will limit the frequency of operation to lower microwave frequency applications. This area of interacting silicon also stores a charge that will reduce the switching speed of the device.

As with any device, there are also some good characteristics of the planar construction when used for PIN diodes. One such area is the series resistance *versus* the current characteristics of such a device. This characteristic is very linear when plotted on a log-log axis. The devices also retain this linearity over a wide dynamic range of forward-bias currents. When you put such a characteristic with the low fabrication expense of planar PIN diodes you have a device which finds many applications for high dynamic range variable attenuators.

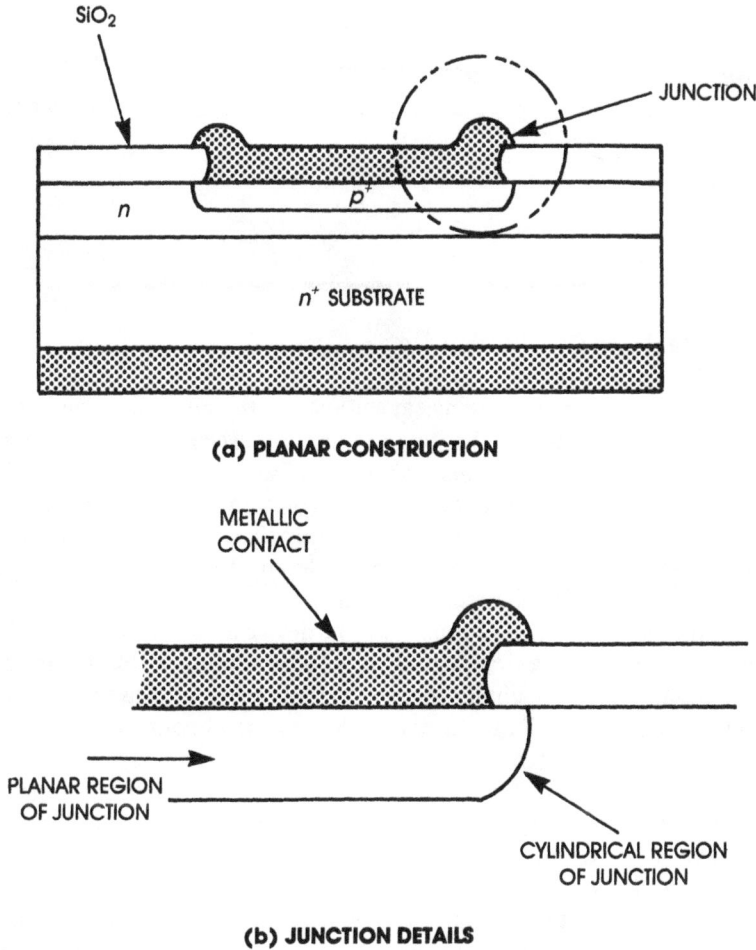

SiO$_2$

JUNCTION

n

p$^+$

n$^+$ SUBSTRATE

(a) PLANAR CONSTRUCTION

METALLIC
CONTACT

PLANAR REGION
OF JUNCTION

CYLINDRICAL REGION
OF JUNCTION

(b) JUNCTION DETAILS

Figure 4.13 Planar Construction

The second type of construction used to fabricate PIN diodes, and probably the most common one, is the *mesa* structure. The basic structure is shown in Figure 4.14. If you look up the word *mesa* in a dictionary and compare that definition to Figure 4.14 you will see where the structure gets its name. The word *mesa* is defined as "a high plateau with steep sides." This, of course, matches the description of the structure we are now characterizing. By having such a structure as a mesa, this means of constructing PIN diodes allows the active intrinsic region very closely to resemble a parallel-plate

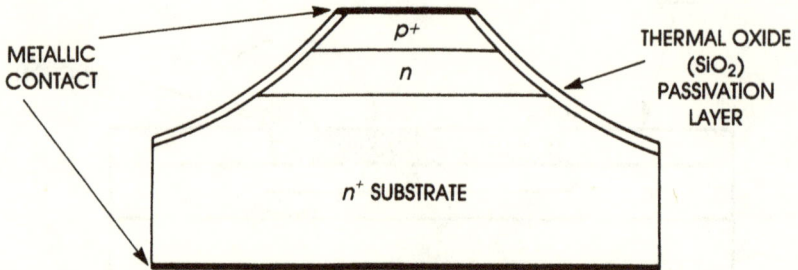

Figure 4.14 Mesa Construction

capacitor with minimum fringing capacitance on the substrate. This becomes evident when you examine the junction area (between the *p* and *n* regions) and discover that there are only planar configurations and no cylindrical regions as were present in the planar construction. The mesa construction also maximizes the surface-breakdown voltage. This results in device-break-down voltages very close to the bulk-breakdown voltage (that is, the voltage breakdown which occurs in pure materials).

A thermal oxide (SiO_2) is used on the device as the surface-passivation layer which ensures maximum reliability. Also adding to the reliability of these structures is the ohmic-contact metallization of titanium tungsten (TiW) and gold (Au). This is a dry refractory metal contact system which is well known for its high reliability.

A combination of the two previous methods of construction (planar and mesa) has been developed by Varian Associates and is appropriately enough called PLESA™. Figure 4.15 shows the process of constructing such a diode as well as the final product. To produce such a PN junction epitaxial diode, an *n* or n^+ epitaxial silicon slice is selected as with both the mesa and planar devices previously discussed. Prior to any diffusions, silicon mesas are formed. Masks are formed for etching by first pyrolytically (a heating process) depositing a silicon nitride (Si_3N_4) layer and an SiO_2 layer on the *n*-layer surface. Suitable etches result in an array of masking dots on the epitaxial silicon as shown in Figure 4.15(a). The mesas are then etched in the silicon slice, and a very dense thermal oxide is grown on the sides. An important consideration at this point is that the PN junctions have not as yet been formed. This allows high-temperature passivation without unwant-ed contaminants. The PLESA™ at this point (Figure 4.15b) is passivated with a window prepared for the diffusion. This die is the same as the planar form except for the geometry of the epitaxial silicon.

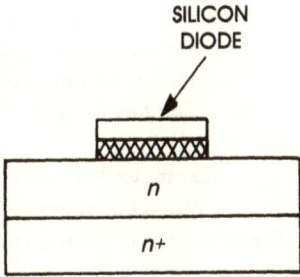

(a) **MESA FORMED**

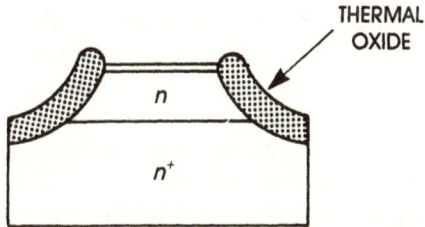

(b) **PASSIVATED WITH A WINDOW**

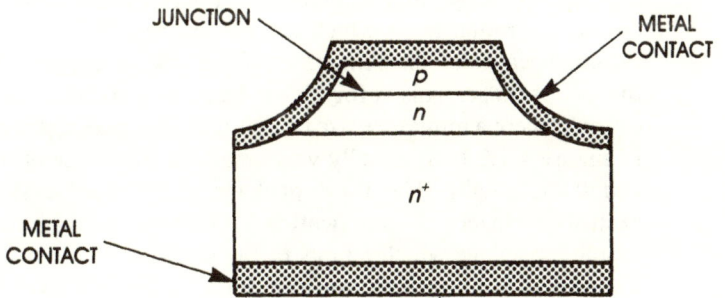

(c) **PLESA™ CONSTRUCTION**

Figure 4.15 PLESA™ Construction

Conventional diffusions through the windows produce junctions under the dense passivation, as in the planar method. Because of the unique processing, the junction is in a mesa geometry resulting in a near parallel-plate capacitance and excellent electrical characteristics. The device is completed, as shown in Figure 4.15(c), by applying final metallization.

The PLESA™ type of construction has actually taken the best characteristics of both the planar and mesa techniques and put them together. This type of device exhibits the good linearity of the planar devices and the low fringing capacitance of the mesa.

The final type of PIN diode construction is also another special combination of existing structures. It is called CERMACHIP™ and is a product of M/A-Com, Inc. This construction is shown in Figure 4.16. The basic geometry of the chip is the same as for the passivated mesa construction. The major difference is in the passivation. In the CERMACHIP™ a thick hard glass passivation is used instead of the thermal oxide of the passivated mesa construction. This glass passivation allows the device to replace a fully packaged hermetically sealed diode chip in many applications. Because of process limitations, this type of device cannot be used for low-level PIN applications. Only chips having large *i* regions (2 mil or greater) and active region diameters greater than 6 mil can be made with this process. Thus, the CERMACHIP™ process is used primarily for high-power applications such as high-power switches and phase shifters. An example would be radar phase array antennas.

Packages available for PIN diodes are shown in Figure 4.17. The first is a glass package used primarily for lower frequency applications. This is due to the radial leads which are not suitable for higher-frequency applications because of the skin effect of microwave energy. The second type is the pill package. This is for higher-frequency stripline or microstrip applications where a miniature size is required. This package is not shown to scale in Figure 4.17. It is actually very small (in the order of 0.020″ in diameter and 0.050″ high). The third package is the double stud which is used exclusively in coaxial applications. The final package is a very common stripline package which adapts to the stripline medium very readily. Thus, a variety of packages are available for a variety of tasks to be performed.

Applications of PIN diodes generally fall into two classifications: attenuators and switches. The configurations used for both of these applications are usually very similar. Figure 4.18 shows a directional coupler (quadrature hybrid) arrangement that can be used in either case. (Specific circuits and applications will be covered in detail in Chapter 6 of this text. This is only a general introduction to PIN diode applications.) In the figure the circuit has a quadrature hybrid at both the input and output. This gives

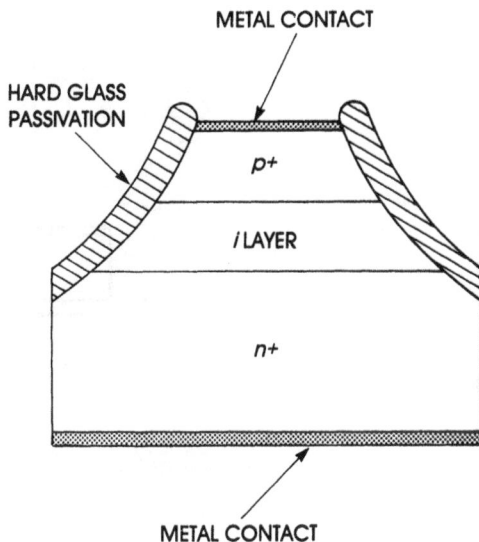

Figure 4.16 CERAMCHIP™ Construction

the circuit a description termed *constant-impedance device*. If we consider the circuit to be an attenuator first, point B in Figure 4.18 has a 50 Ω termination on it. With the PIN diodes forward biased, whatever is applied at the input to the attenuator (IN) will appear at the output (OUT) with a small insertion loss added in. As more and more bias is applied (the diode is being turned off), less power is sent to the output and more is reflected back into the termination at point B. This device is now a continuously variable attenuator. When the diodes are completely turned off (and, in some cases, reverse biased) the maximum attenuation is observed. Thus, we have a continuously variable microwave attenuator with good impedance characteristics.

If we now consider the case where we used a switch, we will take output number 1 at point A in Figure 4.18 and output number 2 at point B (50 Ω resistor is now removed and a terminal, or connector, is attached). If we once again have the diodes forward biased (turned on), the output will appear at point A with a small insertion loss. This is identical to the minimum setting of the continuously variable attenuator discussed above. Now, however, we do not continuously vary the bias but instantaneously switch from forward bias to reverse bias and turn the diode completely off. This causes all of the energy to be reflected back which appears at point B with a small insertion

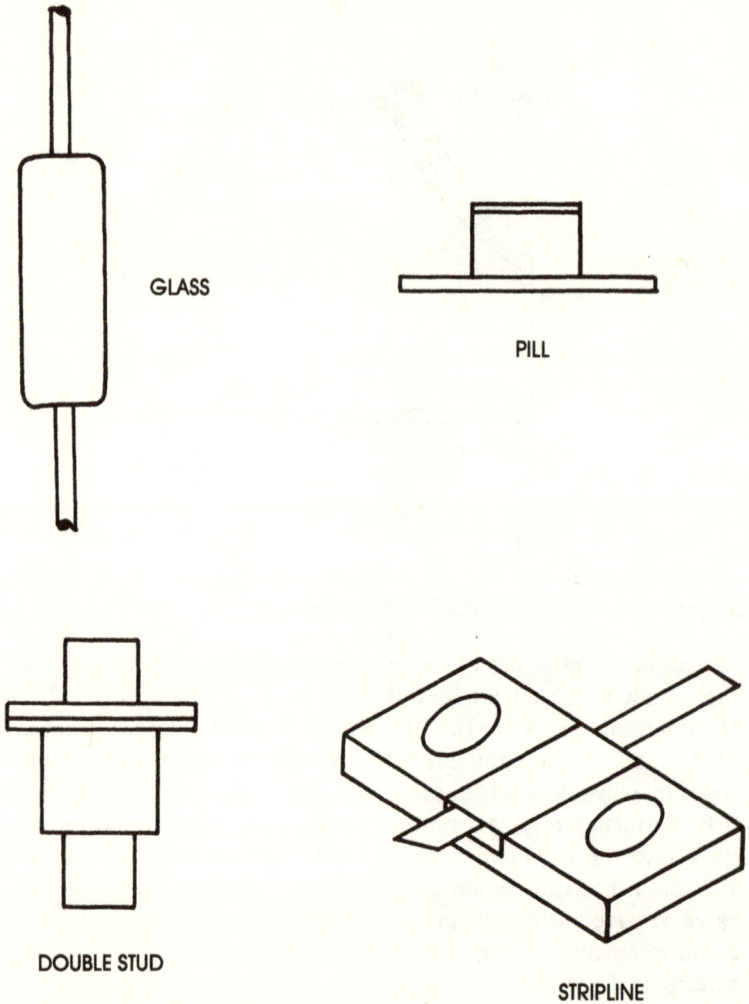

GLASS

PILL

DOUBLE STUD

STRIPLINE

Figure 4.17 PIN Diode Packages

loss. What results is a single pole single throw (SPST) switch which has excellent impedance characteristics, low insertion loss, and an isolation figure which depends on the type and diode arrangement used.

So, we have a general understanding of the operation of PIN diode attenuators and switches. There are many variations of the circuit presented in Figure 4.18. Some of these will be discussed in Chapter 6. For now, suffice it to say that these applications are typical for PIN diodes.

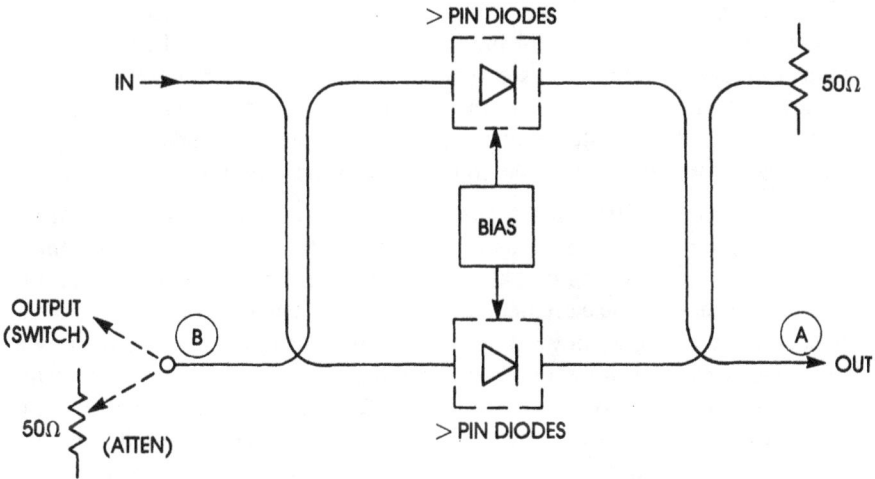

Figure 4.18 PIN Diode Applications

To conclude our discussion of PIN diodes, let us look at five parameters of great importance to PIN diodes, how they would appear on a data sheet, and what they mean. These terms are:

V_B = Breakdown Voltage (V)
$C_j(V)$ = Junction Capacitance (pF)
$R_s(i)$ = Series Resistance (Ω)
τ_L = Carrier Lifetime (ns)
θ_j = Thermal Resistance ($^\circ$C/w)

Breakdown Voltage. The breakdown voltage is controlled by the width of the intrinsic (I) layer. This was discussed previously when covering the different types of PIN diode construction. This number limits the RF voltage swing that may be applied to the diode. If you exceed this voltage, you will be operating basically as simply a PN junction.

Junction Capacitance. This is the junction capacitance of the diode when it is in a fully depleted state. The character, V, in this term is an important part of the parameter. It tells what voltage must be applied to result in the junction capacitance shown on the sheet. For example, C_j (-50) gives the junction capacitance of a device with 50 V applied to deplete the i layer. The voltage is usually more than enough to deplete the layer. When choosing a PIN diode be sure to check both C_j and the voltage needed to obtain this capacitance.

Series Resistance. This is the total resistance of the diode when a certain value of current is flowing through it. Unless otherwise specified, the value is measured at 1 GHz. Just as in the case of C_j (V), R_s (i) may have different values of current depending on the device used. For the values of forward current given, the resistance of the *i* layer is usually very small. The value of R_s (i) is, therefore, very close to minimum series resistance of the diode.

Carrier Lifetime. This is a measure of the ability of a PIN diode to store charge. A pure silicon crystal has a theoretical lifetime of several milliseconds. However, impure doping quickly reduces this lifetime to micro or nanoseconds. The actual measurement of the average carrier lifetime is accomplished by injecting a known amount of charge into the *i* layer of the diode and measuring the amount of time to get it back by using reverse-bias current. The pulse used for this test should have a fast risetime with its pulse width being several times the expected lifetime of the diode being tested. Characterization of the carrier lifetime is of utmost importance when designing switching circuitry.

Thermal Resistance. This is a measure of the ability of the diode to withstand heating effects due to both RF and dc power dissipation. It is defined as the ratio of steady-state temperature rise (°C) of the junction per watt of steady-state power dissipated within it. This parameter is of primary importance when working with high-power switching applications.

So, the PIN diode is now available as a continuously variable or step attenuator at microwave frequencies or as a switching device for a variety of applications. Whichever application you have for your microwave system, the PIN diode will do the job.

4.3 VARACTOR DIODES

The varactor diode is a diode which is used as a variable reactance circuit element. This variable reactance is provided by the junction capacitance (C_j), which varies with applied voltage. With characteristics like this, you can conclude that the varactor diode is a nonlinear device. This non-linearity can produce three different effects. One effect is the switching or modulation of a microwave signal through variations in the reactance of the diode by means of an externally applied bias. This is probably the primary application of such a device.

A second effect is that this nonlinearity causes the generation of harmonies of the applied signal. The third effect is when two microwave signals of different frequency may be applied. The result will be a parametric amplifier or an up-conversion of one of the signals.

The active element of a varactor diode consists of a semiconductor wafer containing a junction, usually formed by diffusion, with the diode generally in a mesa type of construction. Figure 4.19 shows the construction of a typical varactor diode and its equivalent circuit. The components which make up the equivalent circuit, as shown in Figure 4.19(c) are:

C_j = Junction capacitance, which is a function of applied voltage;

R_j = Junction resistance, in parallel with C_j and also is a function of applied voltage;

R_s = Series resistance, may be a function of bias including the resistance of the semiconductor on either side of the junction through which the conduction current passes and the resistance of the ohmic electrical contact to the wafer.

Typical values for these parameters are $C_j(0)$, which is the junction capacitance at zero volts, ranges from 2 pF to 5.5 pF, depending on the application; $R_j(0)$ is greater than 10 megohms; and $R_s(0)$ ranging from 0.45 ohms to 0.7 ohms, once again, depending on the application of the device. Varactor diodes are normally operated under reverse bias, where the junction resistance, normally 10 megohms or more, is negligible in comparison with the microwave capacitive reactance of the junction. Therefore, the equivalent circuit of the reverse-biased varactor waver at microwave frequencies is simply a capacitor (C_j) and resistor (R_s) in series. The equivalent circuit of a forward-biased varactor at microwave frequencies is generally more complicated, since it must include the diffusion capacitance of the injected carriers as well as the effect of these carriers on the conductance of the semiconductor material. (The conductive effect of injected carriers is known as *conductivity modulation*.

The more widely used varactor diode is many times not the wafer (or chip) form of the device. It is this wafer inside a package. Figure 4.20 shows the equivalent circuit of a packaged diode. The term C_c is the *case capacitance*; C_f is the *fringing capacitance* also due to the case; L_s is the *lead* or *series inductance*; and C_j and R_s are the junction capacitance and series resistance which were discussed previously. Notice that Figure 4.20 has no value for the junction resistance, R_j, shown. If you recall, we explained that when the varactor is biased in a zero or reverse-biased condition that R_j was very large (greater than 10 Mohms) compared with the reactance of the junction capacitance at microwave frequencies and was therefore neglected. Thus, the wafer could be represented by a series resistance (R_s) and (C_j).

Although all of the parameters presented above should be considered when choosing a varactor diode, the most important are the junction capacitance, C_j, and its voltage sensitivity; the series resistance, R_s, associated

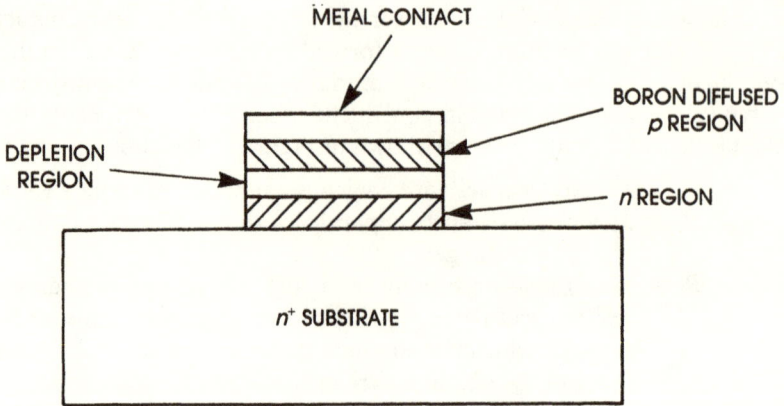

(a) MESA CONSTRUCTION (WAFER ONLY)

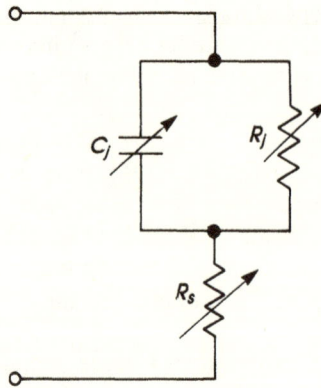

(b) EQUIVALENT CIRCUIT

Figure 4.19 Varactor Diode

with the semiconductor material; and the breakdown voltage. The parameters listed are functions of three conditions:

1. The distribution of the ionized impurities near the PN junction (Is there a large concentration of impurities there or, is it rather small?);
2. The fundamental properties of the semiconductor material (electron and hole mobilities, minority-carrier lifetime, dielectric constant, etc.);
3. The geometry or structure of the junction.

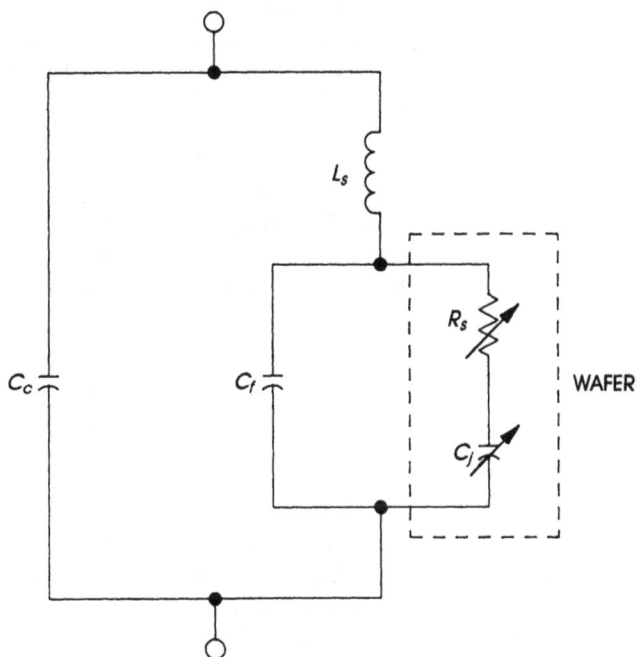

Figure 4.20 Equivalent Circuit of a Packaged Varactor

Careful consideration should be given to the varactor diode you choose for your particular application by evaluating the parameters C, R, and the package characteristics which may limit the frequency of operation.

Typical values for both nonepitaxial and epitaxial varactor diodes are shown in Table 4.2. They are shown for two values of breakdown voltage (10 and 50 V) and two values of C_j (1 and 2.4 pF). The value of junction capacitance is for C_{j0} (zero volts bias). Whenever a subscript B is used it will mean a parameter is at the voltage breakdown point.

Some of the terms presented in Table 4.2 have not been explained as yet. In particular, there are four resistance terms that need to be defined. Figure 4.21 will aid in these definitions. In this figure the following resistances are shown:

R_p = Resistance of the p region
R_n = Resistance of the n region
R_b = Spreading resistance of the substrate
R_c = Total resistance across the wafer

Table 4.2

Parameter	Nonepitaxial		Epitaxial	
V_s (V)	10	50	10	50
C_{j0} (pF)	1	2.4	1	2.4
C_{jb} (pF)	0.27	0.27	0.27	0.27
R_p (Ω)	0.04	0.005	0.04	0.005
R_n (Ω)	0.25	2.1	0.25	2.1
R_b (Ω)	10.6	28.1	0.15	0.056
R_c (Ω)	0.34	0.13	0.34	0.13
R_{total} (Ω)	11.2	30.3	0.78	2.29
f_{CO} (GHz)	14	2.2	204	29
f_{CB} (GHz)	53	19	756	257

You will notice from Table 4.2 that most of the parameters follow one another for nonepitaxial and epitaxial structure. The major difference arises when we encounter R_b. The values for the nonepitaxial structure are many times greater than those in the epitaxial structures. This makes R_b a very dominant parameter in this type of varactor diode. Another major difference between the nonepitaxial and epitaxial structures is the cutoff frequency. You will notice that the cutoff frequencies for the nonepitaxial case is at 53 GHz maximum, while the maximum for the epitaxial case is 756 GHz. These are significant numbers and should be considered very carefully when determining which varactor diode to use. One further point to be considered is that higher cutoff frequencies, either f_{co} or f_{cB}, are always associated with lower breakdown voltages. So you may need to make a tradeoff to ensure that your circuit will function properly.

Varactor diodes are useful elements in microwave circuits because, as previously mentioned, they exhibit nonlinear capacitance and have low loss. Some of the applications of the varactor are in low-noise parametric amplifiers, harmonic-frequency generators, frequency converters, mixers, and the application shown in Figure 4.22, an FM modulator circuit. In this circuit the varactor diodes are used as capacitance-variable components which change the resonant frequency of the original circuit, C_o and L. This is achieved by using the modulating signal (f_m) to vary the voltage of the capacitance of the varactor. When this is combined with the original carrier, the resulting FM segment is very clean and controlled. This concept can also be used to produce a scanning receiver using varactor diodes to tune across a band by application of a sawtooth or triangular wave which will vary the amount of voltage and, thus, the capacitance of the diode.

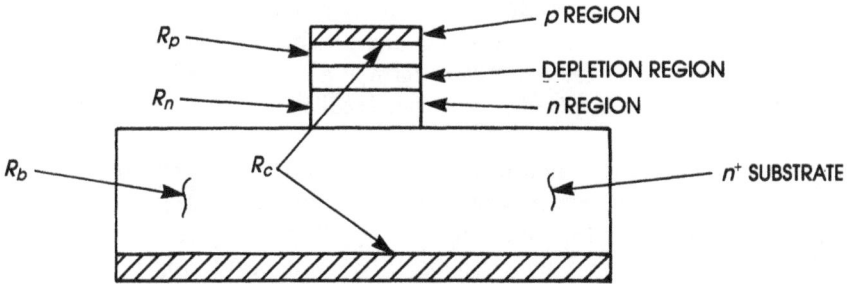

Figure 4.21 Resistance Within a Varactor

Thus the varactor diode can be used in many applications where tuning is required or it is advantageous to use a nonlinear device to accomplish a specific condition.

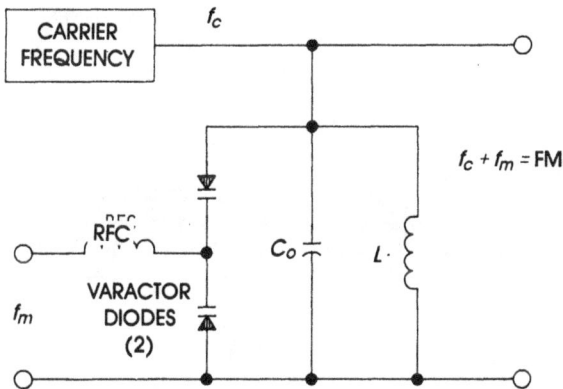

Figure 4.22 FM Modulator

4.4 GUNN DIODES

In the early 1980s it was said that as long as the properties of semi-conductors depended on junctions and as long as the junctions must be made thinner as the frequencies increased, high powers using semiconductor devices at microwave frequencies would not be possible. Devices have been developed, however, which do not depend on a junction for operation. One such class of device exhibits microwave power properties that depend on the behavior of *bulk* semiconductors, rather than junctions. The *Gunn effect* is the main representative of this group and the *Gunn diode* is the main device. The term *diode* may be a confusing term here, since there is no junction and no rectification is involved. The device is called a diode only because it has two terminals.

In 1963, J. B. Gunn discovered the *transferred electron effect*, now known as the Gunn effect. The effect was found to be exhibited by gallium arsenide (GaAs) and indium phosphide (InP). Subsequently, cadmium telluride (CdTe) and indium arsenide (InAs) were also found to possess this property. If a relatively small dc voltage is placed across a thin slice of GaAs, then a negative resistance will be present under certain conditions. The one primary condition is that the voltage gradient across the GaAs slice exceed 3300 V/cm. A GaAs Gunn slice is shown in Figure 4.23.

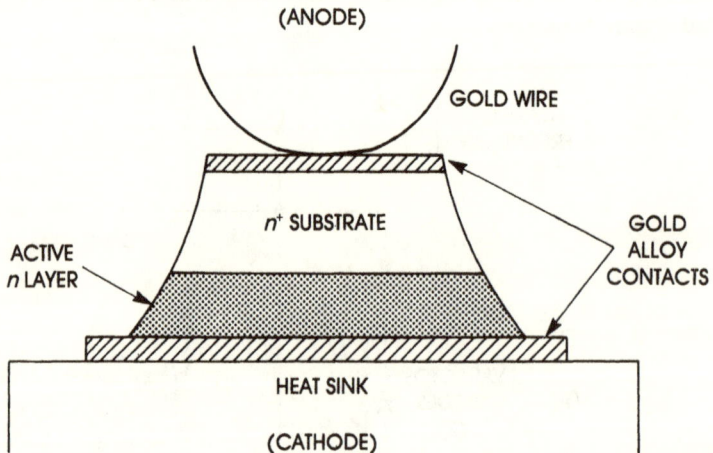

Figure 4.23 GaAs Gunn Slice

As previously stated, the Gunn effect is a *bulk* property of semiconductors and does not depend on either junction or contact properties for its operation. The effect is independent of total voltage or current and is not affected by magnetic fields. Furthermore, it occurs in *n*-type material only, so that it must be associated with electrons rather than holes. The frequency of oscillation produced corresponds closely to the time electrons take to cross such a slice of *n*-type material (such as in Figure 4.23) as a result of the voltage applied. This suggests that a bunch of electrons, called a *domain*, is formed somehow, once per cycle and arrives at the positive end of the slice to excite oscillations in the associated tuned circuit.

The idea of how this device obtains its negative-resistance characteristics requires further investigation. Gallium arsenide is one of a fairly small
number of semiconductor materials which, in an *n*-doped sample, has an
empty energy band higher in energy than the highest filled band. Also, the
size of the forbidden energy gap between these two is relatively small. When
a voltage is applied across a GaAs slice which is *n* type, these electrons flow
as current toward the positive end of the slice. The greater the potential across
the slice, the higher the velocity with which electrons move toward the
positive end and, thus, the greater the current. Thus far this appears much
the same as any positive-resistance device. In this case, however, there is so
much energy imparted to the electrons by the extremely high voltage gradient
(positive resistance devices will never have a gradient anywhere near this
high), that instead of traveling faster and increasing the current, they actually
slow down. This is because the electrons have acquired enough energy to
be transferred to the higher band, which is normally empty. This process
produces an effect called the *transferred electron effect*. Electrons have thus
been transferred from the conduction band to a higher energy band in which
they are much less mobile. Thus the current has been reduced as a result
of a voltage rise. This phenomenon, which is a classic representation of
negative resistance, can be seen in the *I-V* curve for a typical Gunn diode
in Figure 4.24. It can be seen from the figure that as the votlage rises past
what is termed the *threshold negative resistance value*, the current also rises
until this point and then falls. If the bias voltage was to be increased further
and further, this phenomenon would cease, because eventually the voltage
would be sufficient to remove the electron from the higher-energy low-
mobility band, and the current would once again increase. For this reason
the ideal area to operate is in the linear region just past the threshold value
(the area a-b in Figure 4.24). When the Gunn diode is biased in this negative-
resistance region, the field distribution within the sample rearranges itself
to form a thin, high-field domain with the rest of the sample experiencing
a subthreshold field. The domain drifts through the sample at the saturated
drift velocity and collapses when it reaches the anode. The bias current then
momentarily increases until a new domain (or bunch of current carriers)
forms at the cathode. This periodic current increase results in high frequency
power being made available to the external circuit. The output frequency,
thus, is determined by transit time effect (the time it takes for one bunch
to get from the cathode to the anode), but in practice to vary the frequency
over about one octave is possible by tuning the external resonant circuit
(because of modulation of the bias voltage by the RF voltage developed in
the resonant circuit, which can cause the instantaneous diode voltage to fall

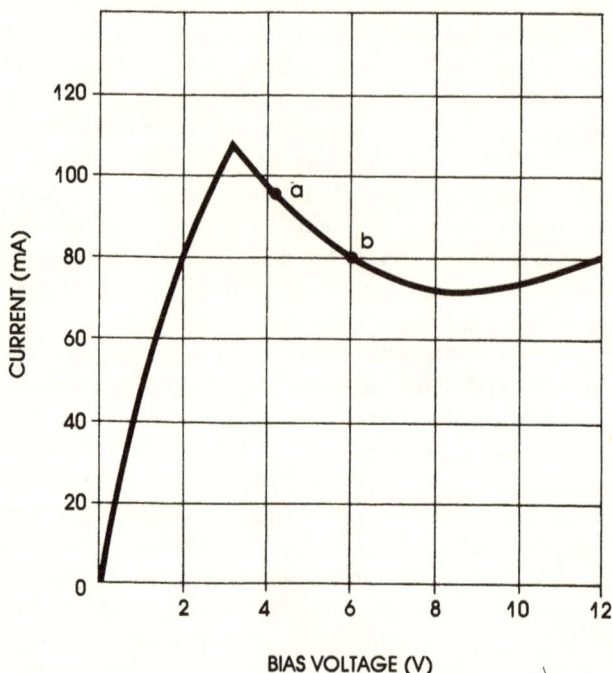

Figure 4.24 *I-V* Curve for a Gunn Diode

below the diode threshold). Formation of subsequent domains is then delayed, giving rise to the name *delayed domain* for the mode of operation.

Limited space-charge accumulation (LSA) operation in Gunn diodes is obtained by tuning the diode in a high-impedance circuit at a frequency such that a full domain cannot form during the period of an oscillation. The RF swing necessary is such that the pulsed operation is normally used. The principal advantage of this mode is that since the material is displaying bulk-negative resistance, the dimensions of the diode are no longer limited by transit-time conderations, meaning that millimeter-wave frequencies can be achieved without the use of impossibly small active regions. A disadvantage of LSA operation is that if domains do start to form, the RF voltages involved are so high that the domain fields may cause an avalanche breakdown, leading to the destruction of the device. Since domains can be created by even the slightest disruption in the consistency of the material, the requirements of the material for LSA operations are extremely strict. Also, it is possible to use Gunn diodes in other modes (quenched domain, multiple domain, and hybrid, for example), but these suffer from some or all of the disadvantages of being little understood, being difficult to control, or requiring diodes of special design.

Certain precautions should be observed when operating Gunn diodes. They should be operated from a constant-voltage power supply. The required operating voltage range is determined by the threshold field, power dissipation, and breakdown considerations. The data sheets should be very carefully checked for maximum voltage ratings. Although in some circumstances a greater output is possible to obtain by operating at higher voltages (pulsed conditions), no guarantee of reliability is given. The quoted ratings are maximum values, and the power supply should be free of any overvoltage transients (which are most likely to occur during switch on). Bias polarity must, of course, also be heeded very carefully. Since the frequency-pushing characteristics of these devices can be as high as 30 MHz/V, noise and ripple on the power supply can produce a significant amount of FM noise on the oscillator output. A well-stabilized power supply is, therefore, necessary to reduce frequency-noise modulation.

Bias-circuit oscillations may occur in certain circumstances due to a slightly negative resistance in the diode dc characteristics just above threshold, and they are found generally in the range of 1 to 100 MHz. Although they are not damaging to the diode, they can result in a poor output spectrum. These oscillations can be suppressed by connecting a $0.01\mu F$ to a $1.0\mu F$ capacitor across the oscillator bias terminals.

The InP (indium phosphide) Gunn device is developed specifically for low-noise amplifiers from 26 GHz to 100 GHz and for medium-power low-noise oscillators from 40 GHz to 140 GHz. Typical oscillator powers so far obtained are shown in Table 4.3.

Table 4.3

Frequency (GHz)	Power Output (mW)
56	200
90	100
95	68
100	44
55 (pulsed, 10% duty)	860

Indium phosphide (InP) has advantages over gallium arsenide (GaAs) for high-frequency millimeter-wave oscillator applications. Two of these advantages are:

• Higher peak-to-valley ratio;
• Reduced scattering time due to a high-threshold field

Peak-to-valley ratio is illustrated in Figure 4.25. Figure 4.25 is a plot of particle velocity *versus* electric field for *n*-type GaAs and InP. InP has a peak(maximum)-to-valley(minimum) ratio of approximately 3, and the GaAs shows less than 2.4. This ratio must be high because the basic efficiency of a transferred-electron oscillator (Gunn device) is strongly influenced by this ratio of the material used. This InP ratio has been shown to be higher than that in GaAs; but even better is the fact that when the materials are heated, the ratio in GaAs is significantly reduced, and the InP reduces both peak-and-valley velocities of nearly the same amounts, preserving the high resultant ratio.

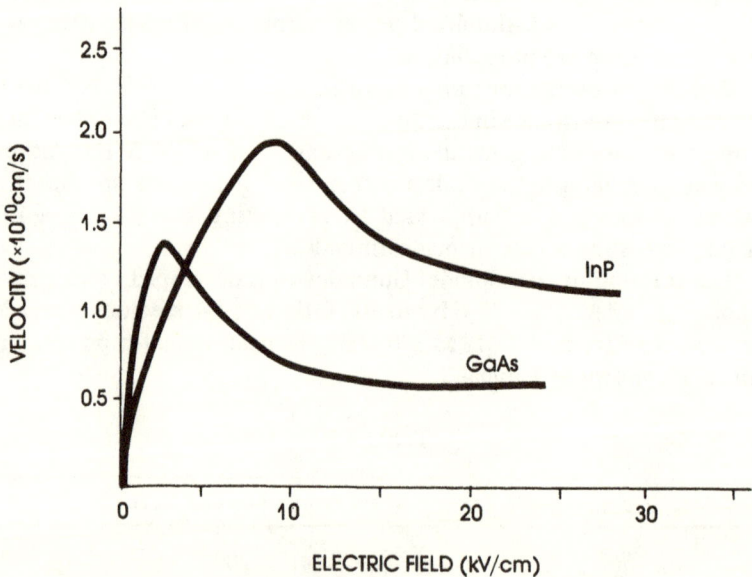

Figure 4.25 Peak-to-Valley Comparison

The scattering process defines the movement of charges (or particles) from one point to another. Each of the scattering processes within the Gunn device takes a certain amount of time to move between these points. (The points referred to are termed the central valley, which is the main valley of the curves, and the satellite valley, which is a secondary valley.) Times considered are those required to move from the central valley to the satellite valley and the reverse time. These times have been shown to be a factor of two smaller in InP than in GaAs. Thus, if we have a GaAs Gunn device operating at 90 GHz, we can expect similar performance of InP at 180 GHz, which is a sizeable improvement in overall performance.

A third characteristic that makes InP more desirable at millimeter frequencies should also be mentioned. That characteristic is the higher impedance on InP. The design of millimeter-wave oscillators is significantly simplified when a large-device negative resistance and small reactance is present, which is the case for InP Gunn devices. The obvious higher-peak velocity of InP (shown in Figure 4.25) permits an increased active layer length, maintaining a transit-time relationship with a low bias-to-threshold ratio.

As a final note it should be pointed out that the wideband negative-resistance characteristics of an InP Gunn device, coupled with its low-noise properties, also make the device an excellent choice as a reflection amplifier. The applications are suitable for millimeter-wave narrow and broadband circuits. To sum up the Gunn device applications we can say,

- Microwave application — GaAs,
- Millimeter application — InP.

4.5 IMPATT DIODES

The name IMPATT stands for *IMPact Avalanche and Transit Time* diode. Its name and what it stands for will become more apparent as our discussions are presented. In our previous discussions we have said that the Gunn diode was a negative resistance device. It was said to have a dynamic dc-negative resistance. This meant that, over a certain range, current decreased with an increase in voltage, and *vice versa*. This particular point was pursued no further, it being taken for granted that any device which exhibits a dynamic negative resistance for direct current will also exhibit it for alternating current. That is to say, if an alternating voltage is applied, current will rise when voltage falls, at an ac rate. We may thus now redefine negative resistance as that property of a device which causes the current through it to be 180° out of phase with the voltage across it. The point is important here, because this is the only kind of negative resistance exhibited by the IMPATT diode.

A combination of delay involved in generating avalanche current multiplication, together with delay due to transit time through a drift space, provides the necessary 180° phase difference between applied voltage and the resulting current in an IMPATT diode. The cross section of the active region of this device is shown in Figure 4.26. Note that it is a diode, the junction being between the *p* and *n* layers.

An extremely high-voltage gradient is applied to the IMPATT diode, of the order of 400 kV/cm, eventually resulting in a very high current. A normal diode would very quickly break down under these conditions, but

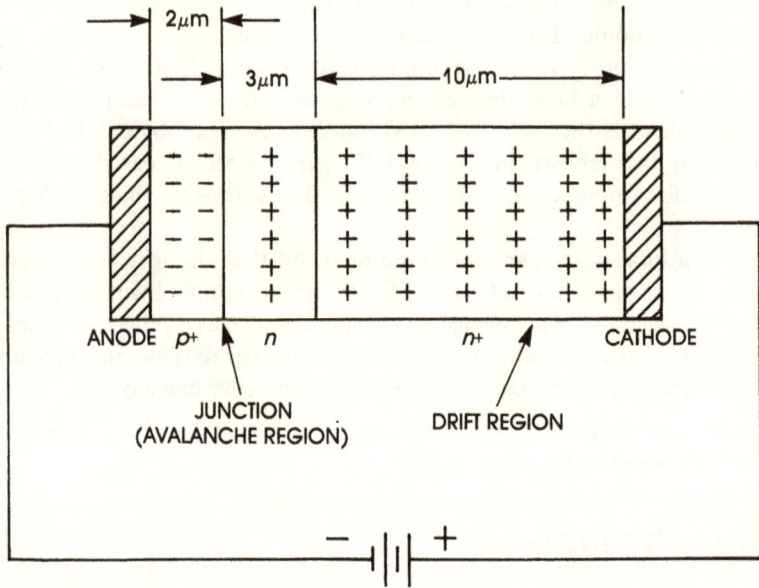

Figure 4.26 IMPATT Schematic

the IMPATT diode is constructed so as to be able to withstand these conditions repeatedly. Such a high potential gradient, back-biasing the diode, causes a flow of minority carriers across the junction. If it is now assumed that oscillations exist, we may consider the effect of a positive swing of the RF voltage superimposed on top of the high-dc voltage. Electron and hole velocity has now become so high that these carriers form additional holes and electrons by knocking them out of the crystal structure, by so-called impact ionization. These additional carriers continue the process at the junction, and it now snowballs into an avalanche. If the original dc field was just at the threshold of allowing this situation to develop, this voltage will be exceeded during the whole of the RF positive cycle, and avalanche current multiplication will be taking place during this entire time. Since it is a multiplication process, however, avalanche is not instantaneous. Indeed, as shown in Figure 4.27, the process takes a time such that the current-pulse maximum, at the junction, occurs at the instant when the RF voltage across the diode is zero and going negative. A 90° phase difference between voltage and current has thus been obtained.

As so far described, the current pulse in the IMPATT diode is situated at the junction. It does not however, stay there. Because of the reverse bias, the current pulse flows to the cathode, at a drift velocity dependent on the

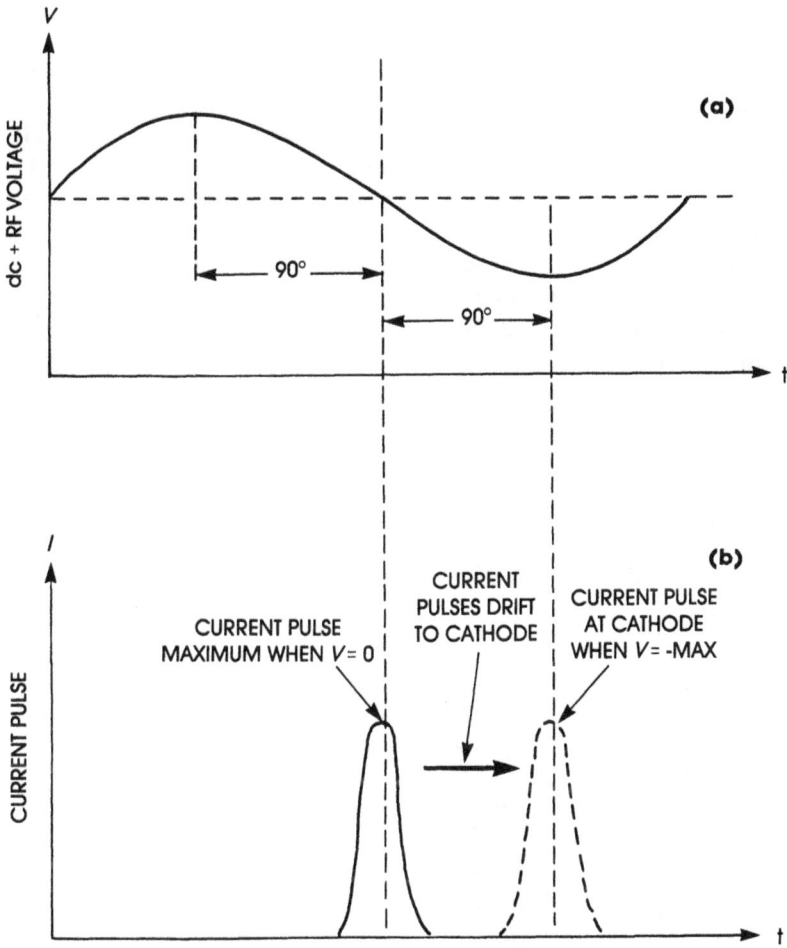

Figure 4.27 IMPATT Behavior

presence of the high-dc field. The time taken by the pulse to reach the cathode depends on this velocity, and of course on the thickness of the highly doped (*n*) layer. Also, the thickness of the drift region is selected so that the time taken for the current pulse to arrive at the cathode corresponds to a further 90° phase difference. Thus, as shown in Figure 4.27, when the current pulse actually arrives at the cathode terminal, the RF voltage there is at its negative peak. Accordingly, voltage and current in the IMPATT diode are 180° out of phase, and a dynamic RF negative resistance has been proved to exist. As with other devices studied in this chapter, such a negative resistance lends itself to use in oscillators or amplifiers. Because of the short times involved, these can be microwave.

Commercial IMPATT diodes have been available for quite some time. They are made of either silicon or gallium arsenide, invariably epitaxial and mostly mesa; some have Schottky-barrier junctions. Gallium arsenide is technically preferable. It gives lower noise, higher efficiencies, and higher maximum operating frequencies. It is also more difficult to work, however, and more expensive. Silicon diodes came first, and even now give higher-output powers than GaAs for commercial diodes. It is expected that silicon diodes will continue to be used for less exacting applications.

The IMPATT diode shown in Figure 4.28 is typical of what is man-ufactured and could house a chip made of either GaAs or silicon. The construction is deceptively simple. A lot of thought and development have gone into its manufacture, however, particularly the contacts, which must have extremely low ohmic and thermal resistance. In a practical circuit, the IMPATT diode is generally embedded in the wall of a cavity, which then acts as an external heat sink.

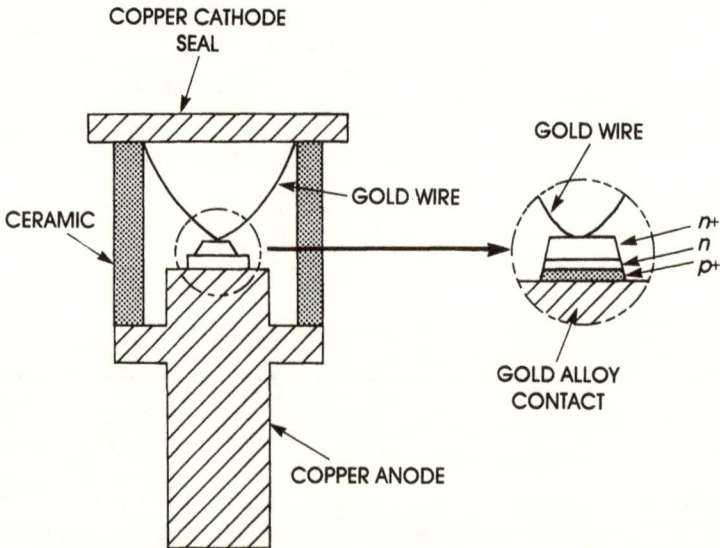

Figure 4.28 IMPATT Diode

It should be noted that, until recently, practical IMPATT diodes were unlike Read's original proposal. This called for a double drift region, whereas Figures 4.26 and 4.28 show diodes with single (*n*) drift regions. The reason for the departure from what was theoretically a high-efficiency structure was difficulty in fabrication. This has more recently been solved, and GaAs RIMPATT (*read*-IMPATT) devices are beginning to show efficiencies approaching 30% in the laboratory. That is more than three times as high as could be expected from a commercial IMPATT diode.

4.6 TRAPATT DIODES

The term TRAPATT stands for *TRApped Plasma Avalanche Triggered Transit* diode. The TRAPATT diode is derived from the IMPATT diode and is closely related to it. Indeed, at first it was merely a different, "anomalous," method of operating the IMPATT diode. A greatly simplified operation will now be described.

Consider an IMPATT diode mounted in a coaxial cavity, so arranged that there is a short circuit a half-wavelength away from the diode at the IMPATT operating frequency. When oscillations begin, most of the power will be reflected across the diode, and thus the RF field across it will be many times the normal value for IMPATT operation. This will rapidly cause the total voltage across the diode to rise well above the breakdown threshold value used in IMPATT operation. As avalanche now takes place, a plasma of generated electrons and holes is generated, placing a large potential across the junction, which opposes the applied dc voltage. The total voltage is thereby reduced, and the current pulse is, as it were, trapped behind it. When this pulse travels across the *n*-drift region of the semiconductor chip, the voltage across it is thus much lower than in IMPATT operation. This has two effects. The first is a much slower drift velocity and, consequently, longer transit time, so that for a given thickness the operating frequency is several times lower than for corresponding IMPATT operation. The second point of great interest it that when the current pulse does arrive at the cathode, the diode voltage is much lower than in an IMPATT diode. Thus dissipation is also much lower, and efficiency much higher. The operation is similar to class C, and indeed the TRAPATT diode lends itself to pulsed instead of CW operation. It must be reiterated that the foregoing is a much simplified description of the operating mechanism. It is, however, adequate to emphasize the similarity to the IMPATT as well as the individual features and applications (mainly pulse operation) which are unique to the TRAPATT.

Commercial TRAPATT diodes are constructed of Si and GaAs with structures corresponding to those of IMPATT diodes, but with gradual, rather than abrupt, changes in doping level between the junction and the anode. Furthermore, they use complementary n^+-p-p^+ structures as shown in Figure 4.29 instead of the p^+-n-n^+ IMPATT chip of Figure 4.26 for reasons

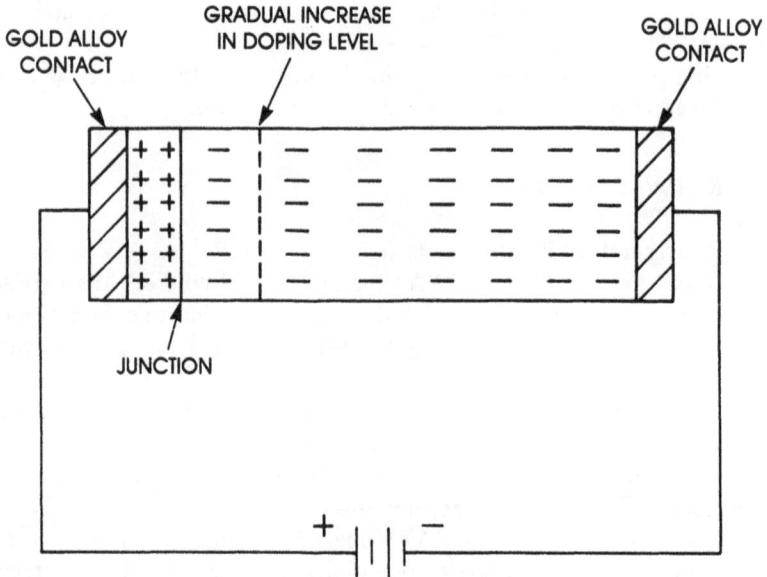

Figure 4.29 TRAPATT Diode

of better dissipation. The two figures should be examined in conjunction with each other.

Because the drift velocity in a TRAPATT diode is much less than in an IMPATT diode, either operating frequencies must be lower or the active regions must be made thinner. In fact, both these considerations are borne out by results thus far obtained. On the one hand, most good TRAPATT results have been for frequencies under 8 GHz. On the other hand, it has been found that by the time 5 GHz is reached, the width of the depletion layer is only 2μm, which is just about the lower practical limit. Since the TRAPATT pulse is rich in harmonics, however, amplifiers or oscillators can be designed to tune to these harmonics, and operation above X band in this manner is possible.

4.7 SUMMARY

The preceding chapter has shown, if nothing else, that the microwave diode is much more than a two-element device used to rectify an ac signal. We have seen how these tiny devices can be constructed so as to mix, detect, attenuate, switch, amplify, oscillate, and modulate microwave signals with great repeatability, reliability, and efficiency. From the general purpose

Schottky diode to the Gunn "diode" which does not even have a junction, to the IMPATTs and TRAPATTs which are the devices of the 1980s and beyond, each has a definite place in this ever-growing field of microwaves.

Chapter 5
Microwave Transistors

The microwave transistor is the device which has probably advanced the most of any microwave device over the years. This statement may be challenged after reading the previous chapter on diodes, but, even though many new diodes have been introduced over a period of time, the transistor is the device which has advanced and improved on its basic designs and structure while introducing basically only two new types.

To aid in understanding just how far the microwave transistor has advanced we will give a brief history of the transistor (*transfer resistor*). The first transistors were fabricated by W. H. Brattain and J. Bardeen in 1948, while they were studying the properties of germanium semiconductor rectifiers at Bell Laboratories. During these studies they observed that the flow of current through the rectifier could be controlled if a third electrode was added to the device. This electrode would serve much the same purpose that the control grid was then doing in triode vacuum tubes. What resulted from this observation was the *point-contact transistor*. This structure is illlustrated in Figure 5.1. To fabricate such a device the collector and emitter wires are pressed into an *n*-type germanium material similar to the method employed in early crystal sets which used "cat's whiskers." This arrangement is the basic germanium rectifier that Brattain and Bardeen were investigating when they discovered the transistor characteristics. With this basic rectifier configuration established, the *p* regions are introduced into the device by a process called *forming*. This process involves the passage of a large current through the wire and into the germanium for a short period of time. The *p*-type germanium forms almost immediately at the contact point as a result of the diffusion of some acceptor impurities from the wire into the germanium. Thus, the name *point-contact transistor* is derived and the resulting device is a *pnp* transistor.

This device, of course, was a major breakthrough for the electronics industry, because it promised to replace many of the bulky, hot, and high power consuming vacuum tubes of the day. The point contact transistor, however, had drawbacks of its own. This revolutionary device generated

EMITTER   COLLECTOR

p   p

n

BASE

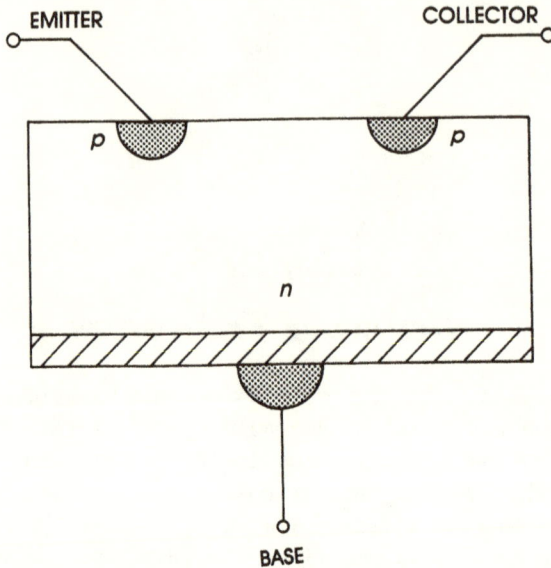

Figure 5.1 Point Contact Transistor

much more noise internally than the vacuum tube it was designed to replace; since it was not hermetically sealed, it could not tolerate temperature or humidity; and it was a very fragile device to handle. These problems were rather serious if one was to use the transistor for any useful and demanding applications other than sitting on a bench and controlling the flow of current. This came about in 1949 when W. Schockley published a paper describing the possibilities of producing a *junction transistor*. This concept today is one that is widely known and used. These devices, as proposed, were much more rugged and had much improved noise performance.

The first of the junction transistors were produced using a process called a *grown-junction technique*. The term, *to grow*, is one which is very common in semiconductor discussions, and, as we can see, it is one which goes back to the earliest days of transistors. The *growing* of junctions is accomplished with what is termed a *seed* of material (this is a small quantity of material used as a foundation for the finished product). The seed is touched to the surface of a bath or *melt* of molten semiconductor and is slowly pulled away. (The semiconductor is germanium, silicon, etc.) The size of the seed is increased, or grows, because the melt in the area of the seed adheres to the seed and effectively "freezes" when withdrawn from the melt solution. Characteristics of the solution can be changed simply by adding more impurities to the melt or by changing the impurities which are added. When the crystals are completely "grown," they are cut to the appropriate size and leads are attached to the proper areas. These devices are also called *double-doped transistors*.

   Alloy or *fused-junction* devices were the next to appear on the transistor scene in 1951. These turned out to be primarily *pnp*-type transistors, whereas the grown junction was primarily *npn*. A wafer of base material (usually *n*-type) germanium is held in a jig between two pellets of impurity (possibly indium). The structure is held until the impurity melts (155°C for indium) and the molten impurity penetrates the base material. During cooling, crystal growth results in a completed three-layer structure. This is shown in Figure 5.2. This device is characterized by a very high junction capacitance due to

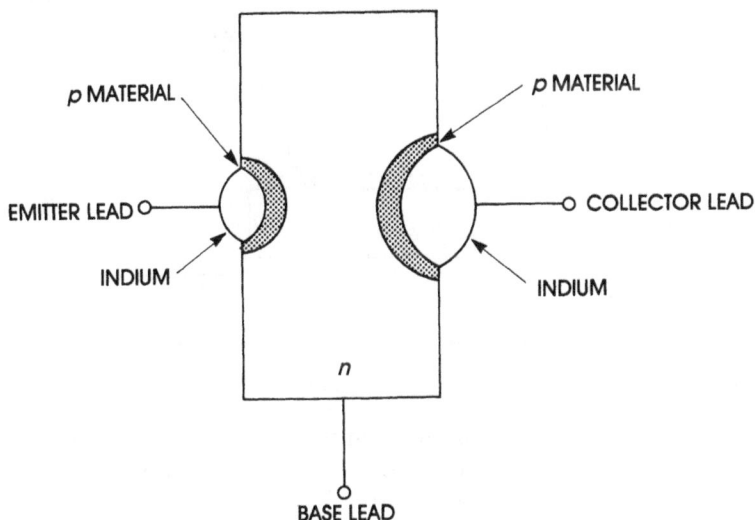

Figure 5.2 Alloy Junction Transistor

the fusing process during construction. Thus, this type of device is not used for high-frequency applications.

   The disadvantage, and thus the limitation, of the alloy-junction transistor was overcome in 1953 by the Philco Corporation when the *surface-barrier transistor* was developed. This type of transistor still finds many applications in today's electronic circuits. Modifications have been made to the original 1953 version, but the basic device and its principles are still what are needed and used. Construction of this device was achieved by a manufacturing process that subjected a small piece of germanium to two jets of electrolyte, such as a solution of indium chloride, which strike the opposite faces of the germanium-base slab. Etching of the semiconductor was achieved by passing a current through the electrolyte stream and the germanium. When the desired base thickness was achieved, current polarities were reversed, and the jets plated the metal emitter and collector contacts on either side of the base. Improvements were made to this procedure by the addition of impurity diffusion techniques. In this process, impurities of one conductivity type are made to diffuse at high temperatures into the surface of a bar of the opposite type. This results in the formation of a junction we discussed in detail in Chapter 3, the PN junction.

The next generation of transistor was the *mesa* or *diffused-base transistor*. In this device phosphorus pentoxide (P_2O_5) may be diffused into an

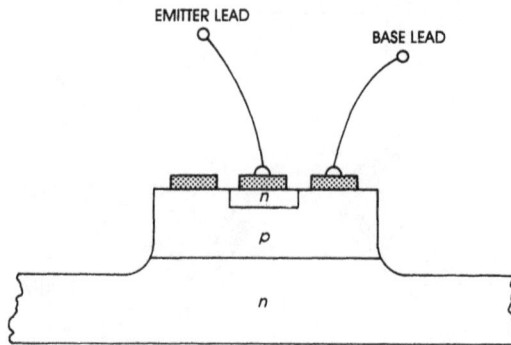

Figure 5.3 Mesa Transistor

n-type silicon collector region to form the base of the device. Acid etching of the proper areas results in a plateau or mesa. You will recall this type of structure from Chapter 4. Figure 5.3 shows how this type of structure applies to transistors.

The final two types of transistors are the *planar* and *epitaxial*. These are the transistors used in microwaves today. The planar construction was shown in Chapter 4 when we discussed PIN diodes and, as you will recall, uses diffusion and surface passivation that protects the surfaces and junction edges from contamination. Leakage occurs in this type of construction because silicon dioxide (SiO_2), which is an insulator, is formed on the exposed surfaces. You will recognize the planar construction once again as it is shown for a transistor in Figure 5.4.

The *epitaxial* transistor is similar to the planar construction, except that it has a thin layer of low-conductivity material for its collector, with the remainder of the collector region of high conductivity. It is for this reason

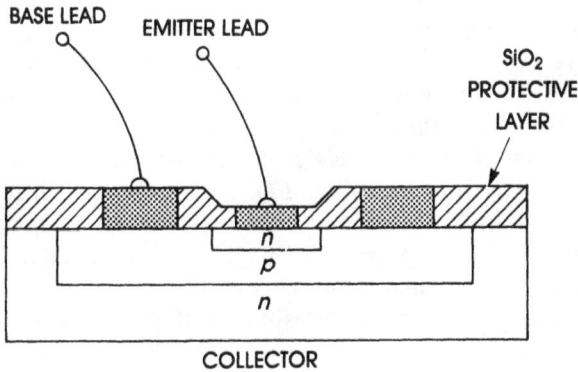

Figure 5.4 Planar Construction

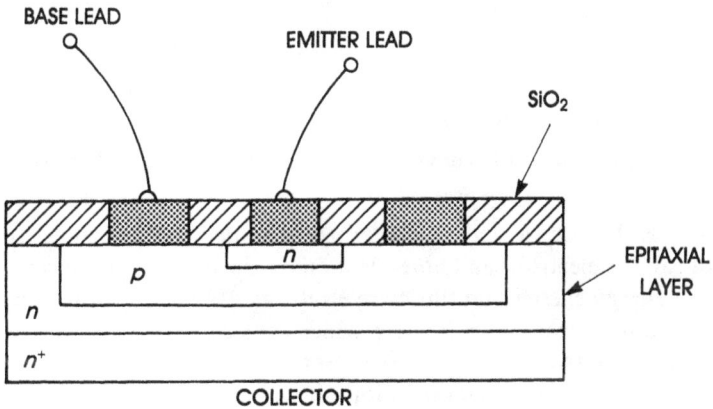

Figure 5.5 Epitaxial Construction

that an epitaxial device is sometimes referred to as an epitaxial-collector type of construction. This type of construction, as we have previously stated, is the main type used for microwave bipolar transistors today. Its construction is shown in Figure 5.5.

The next phase of our history lesson involves the *field-effect transistor (FET)*. Although it was known in theory as far back as 1926 and analyzed completely by Schockley in 1952, the FET was not successfully fabricated for general usage until 1961. Beyond this, it was not used for microwave applications until the mid-1970s, when gallium arsenide pushed the frequency operating area for microwave solid-state devices well up into the high microwave frequency spectrum. The FET has certain advantages over a junction device. These will be investigated as we discuss both the bipolar and FET in detail in the following sections.

The final chapter, to this point, in the transistor story is the *high electron mobility transistor (HEMT)*. This device, which found its way to the microwave industry around 1984, extends the frequency range even higher due to its main property — high electron mobility. This device will be discussed in detail later in this chapter.

With the preceding history lesson and the previous chapter on junction theory, you are now ready to investigate the three types of transistor presented: *bipolar*, *FET*, and *HEMT*.

5.1 BIPOLAR TRANSISTORS

If you were to look up the term *bipolar* in the dictionary you would find it would say it was "having two poles, or extremities." The dictionary of electronics defines it as:

Having to do with a device in which both majority and minority carriers are present.

This is the one we will be concerned with, because it involves the two types of carrier, electrons and holes. With most devices we are concerned only with one energy carrier and those are electrons. As discussed in Chapter 3 in PN junctions, however, the bipolar, which consists of two PN junctions, operates because of two types of carrier. Both electron and hole movement are of vital importance in the operation of the bipolar transistor.

The bipolar transistor finds many applications in microwaves, usually at 4 GHz and below. They are used for low-noise devices, low-power linear amplifiers, and higher-power class-C power amplifiers. Each of these classifications; low noise, linear power, and high power, will be discussed in this chapter.

5.1.1 Low-Noise Transistors

The low-noise bipolar transistor finds its most prominent application as an amplifier in the front end of a microwave receiver. In this application a device that will amplify the low-level input signal and introduce a minimum amount of noise in the process is needed. To obtain this low-noise characteristic requires special construction techniques. As we describe this construction, two terms will be coming up, *planar* and *epitaxial*. These terms were mentioned previously in our introduction, but it would be beneficial for us to get a more detailed explanation of them at this point.

Recall, from our previous discussions, the *planar* construction uses diffusion and surface passivation that protects surfaces and junction edges from contamination. This construction makes the protected areas less prone to surface problems that sometimes occur when structures such as the *mesa* construction are used. The term, *planar*, is one that denotes that both emitter-base and base-collector junctions of the transistor intersect the device surface in a common plane. Perhaps a better term for the structure is *co-planar*. The real significance of the *planar* structure, however, is that the fabrication technique of diffusing dopants (or impurities) through an oxide mask results in the junction being formed beneath a protective oxide layer. Thus the previously mentioned isolation from surface problems. It can be seen by examining previous figures (Figure 5.3 for mesa and 5.4 for planar) how the planar provides the protection discussed above.

The term *epitaxial* is actually a shortening of the term epitaxial collector. This term means that the collector region of the transistor is formed by the epitaxial technique rather than by the diffusion method as mentioned earlier for the base and emitter. The epitaxial layer is formed by condensing a single-crystal film of semiconductor material on a wafer of substance, usually the same material. Thus an epitaxial device is one in which the collector region is formed on a low-resistivity silicon substrate. Note in Figure 5.5, which is the epitaxial construction of a transistor, where the epitaxial layer is located. Location is the important factor, not its material. It is actually the same material as the collector.

The base and emitter regions of the epitaxial device are diffused into the EPI layer, as it is termed. This diffusion can be noticed in Figure 5.5, as well as the fact that the epitaxial structure can be thought of as the planar construction with an extra layer that separates the collector from the base and emitter structures. The epitaxial technique lends itself to precise tailoring of the collector-region thickness and resistivity that improve device performance and uniformity.

The internal "geometry" of most low-noise transistors is an interdigital type of construction. This interdigital construction is illustrated in Figure 5.6. This geometry achieves the lowest base resistance without sacrificing gain. This low-base resistance high-gain combination results in a low-noise

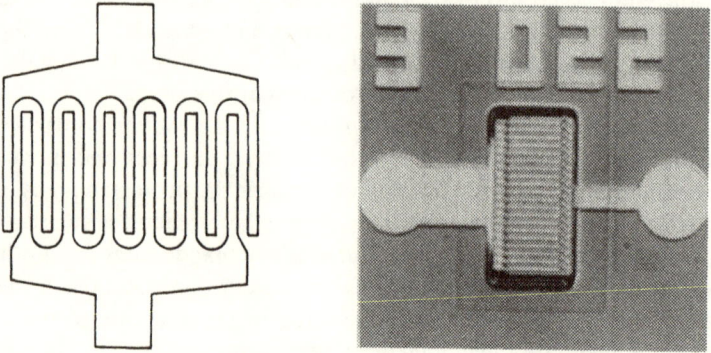

Figure 5.6 Interdigital Construction (Photo Courtesy of Avantek, Inc.)

figure. To understand why this comes about, consider what the noise figure of a transistor is.

The noise figure of a transistor is a measure of how much the signal-to-noise ratio (comparison of signal level to noise level) is degraded as the signal passes through the transistor. The degradation that does occur is due to the presence of three noise sources within the chip itself. These are:

- Shot noise in the emitter-base circuit,
- Shot noise in the collector circuit,
- Thermal noise generated by the base resistance.

Shot noise is the noise generated at a semiconductor junction due to the current that is passing through it. It is directly proportional to the square root of the current applied to the junction. It is, therefore, logical to keep the current through the device as low as possible. Notice that when a low-noise transistor is presented later that the lower noise figures are achieved when the collector current (I_c) is low.

Thermal noise is the noise generated by the random thermal motion of charged particles (or electrons within the base region). Remember that we referred to the fact that the interdigital structure achieved the lowest base resistance without sacrificing gain. If we have a low resistance in the base region, we will have a small voltage dropped across that region and thus lower power dissipated. This lower power will mean very slight thermal activity in the base region and, therefore, low thermal noise.

With the information above, we can say that the best low-noise transistor should have a construction that offers an "efficient" junction structure and a low base resistance. By efficient junction, we mean one that will provide the *pn* junction action with a minimum amount of current or loss across that junction.

How would you make such a low-noise device for microwave operation? If we were transistor designers for a while, we could construct a
representative device. Figure 5.7 shows the step-by-step process.

The process begins in Figure 5.7(a) with an n-type epitaxially grown
silicon layer. This layer ranges in thickness from 2 to 5 μm (0.00007874 in
to 0.00019685 in). This silicon layer is the epitaxial layer shown previously
in Figure 5.7 and is supported by a substrate that is the collector.

A thermally grown oxide layer several thousand angstroms thick (an
angstrom equals 3.937×10^{-9} inches or 1×10^{-10} meters) is formed on the
EPI layer and the base contact cuts are defined and etched by photoresist
techniques as shown in Figure 5.7(b). Through the open areas a heavily doped
p-type (boron) diffusion is made as shown in Figure 5.7(c). Note that these
areas, which are the base region, are of heavy diffusion so that a high
concentration of p-type material will.be available for each finger of the base
construction. This high concentration lowers the metallization contact resistance and provides low-resistance contacts to the transistor base region.
Remember that low base resistance is an important prerequisite for low-noise
devices.

Figure 5.7(d) shows the base area being cut into the previously deposited oxide layer. A precisely controlled amount of lightly doped p-type
diffusions (boron once again) is inserted through the base area opening. This
diffusion now forms the base and is automatically connected to the diffusion
mode in Figure 5.7(c).

With the base of the transistor taken care of, the emitter is next to be
produced. The emitter opening is defined and etched as shown in Figure
5.7(e). This opening must be located precisely between the two previous
diffusions used for the base. The device is now completed by diffusing a
shallow, heavily doped n-type emitter into the emitter opening etched previously, which can be seen in Figure 5.7(f).

To achieve good microwave performance in a device such as the one
just described, the depth of the diffusions is kept very small. The total
junction depth of the base of a 2-GHz transistor is only on the order of 0.3
μm (0.0000118 in). The emitter will penetrate only about 0.2 μm (0.00000787
in) and the base width, the difference between the emitter and base junction
depths, will, therefore, be around 0.1 μm (0.000003937 in).

Contact to the finished transistor is accomplished through metal fingers
that align with the open p-base diffusion and the emitter fingers. These
openings are made in two steps. First, the emitter oxide is "washed" away
by a very short acid etch. Second, the p-base diffusion contact opening is

(a) *N*-TYPE SILICON LAYER   **(b)** OXIDE LAYER FORMED

(c) DIFFUSION OF *P*-TYPE MATERIAL **(d)** BASE FORMED AND CONNECTED TO PREVIOUS DIFFUSION OF *P*-TYPE MATERIAL

(e) EMITTER WINDOW ETCHED   **(f)** *N*-TYPE EMITTER DIFFUSED INTO PREVIOUS EMITTER OPENING

(g) PLANE VIEW

(h) CROSS SECTION

KEY

- OXIDE
- LIGHT DIFFUSION
- HEAVY DIFFUSION
- METAL

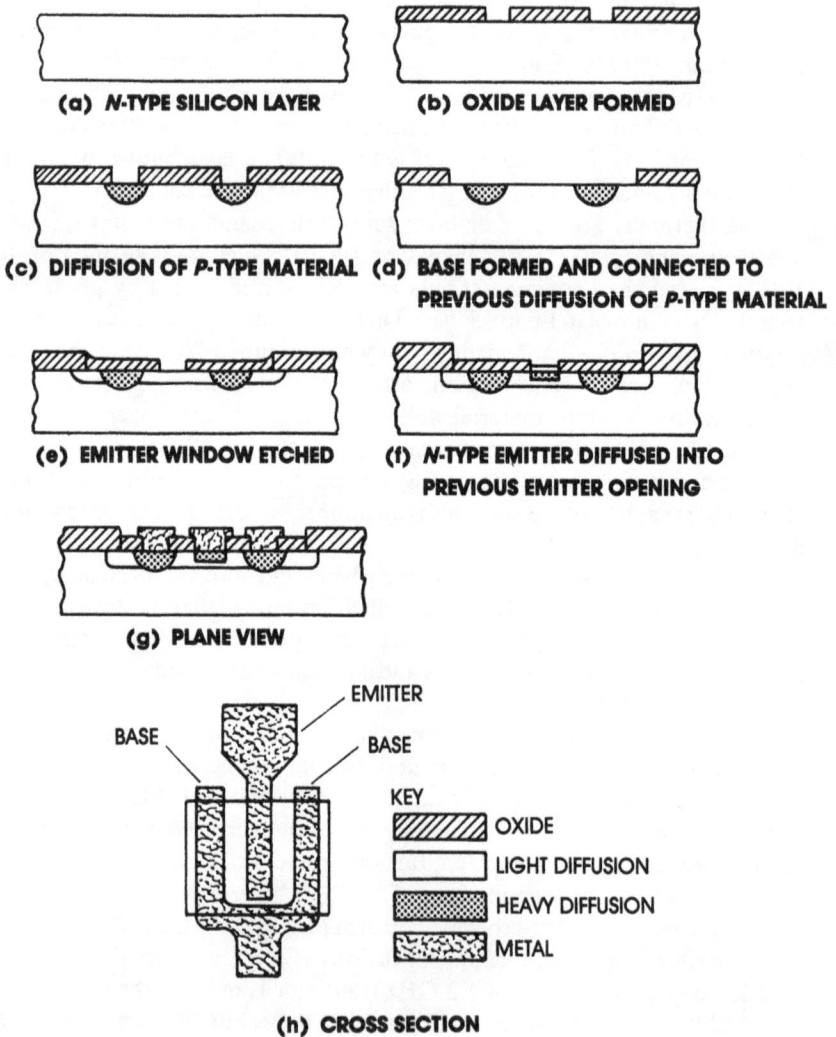

Figure 5.7 Transistor Construction

defined and etched by photoresist techniques. The metallization is deposited in a film over the whole wafer area, and the fingers and bonding pads defined by chemical etching. A final cross section and plane view are shown in Figures 5.7(g) and 5.7(h). The pattern shown in Figure 5.7(h) was also presented in Figure 5.6, which was the internal construction of an actual device.

A more recent method of transistor construction is *ion implantation*. This is a method of embedding the dopant into a wafer by accelerating the ions toward the wafer with a high electric field. The advantage of this process is that both the dopant dose and depth are controlled with much more precision. This precision results in both a much higher yield of devices and superior performance, since the devices can be fabricated closer to the original design.

With the low-noise transistor defined and constructed we will now look at a data sheet which would be associated with such a device. This data sheet is for a general low-noise microwave bipolar transistor. The primary parameters to be concerned with are frequency, noise figure, and gain. Each of these will be discussed below with a variety of gains covered.

The first parameter, *frequency*, should be just that — the first parameter you look at. This is because you may find a transistor with an excellent noise figure, but it only operates to 2 GHz. If your application is from 3.0 to 3.5 GHz, the excellent noise figure will do you no good at all. So the first thing to consider when selecting a transistor is its frequency range. Frequency range is usually one of the first items on the data sheet, and it will appear in an opening paragraph termed "Description" or "Description and Applications." It also may appear in a summary of specifications termed "Features." Wherever it appears, be sure to pick this up first. A paragraph entitled *Description* would read something like this:

> The ABC-442 is a silicon bipolar transistor designed for use in low-noise, small-signal amplifiers up to 4 GHz. This device features excellent gain characteristics while maintaining its specified low noise figure over a broad range of frequencies. It is available in chip form as well as in our 70-mil and 100-mil packages.

Check out this paragraph on the transistor data sheet first. It will save time and aggravation later on.

The next logical parameter to look at on the data sheet is, of course, the *noise figure*. This figure is sometimes called *spot noise figure* or *minimum noise figure*. When reading this number, be sure to read the conditions that go along with it. Simply saying that a transistor has a noise figure of 2.8 dB means little or nothing. If, however, the manufacturer says the device has a noise figure of 2.8 dB at 4 GHz with $V_{CE} = 10$ V and $I_c = 5$ mA, the device is characaterized so that we can determine whether we can use it in a particular application. (V_{CE} is the collector-to-emitter voltage, and I_c is the collector current.) This type of detailed characterization is absolutely necessary, since the design resulting from the use of a particular device depends on the S-parameter data that are taken at specific conditions (V_{CE},

I_c). (S-parameters are the parameters which define the transistor in reference to a 50 Ω system.) To have the proper matching circuits and ensure that the transistor will operate with the expected noise characteristics, these conditions must be duplicated. Figure 5.8 shows the two conditions governing the noise figure of a device. Figure 5.8(a) is noise figure versus frequency and

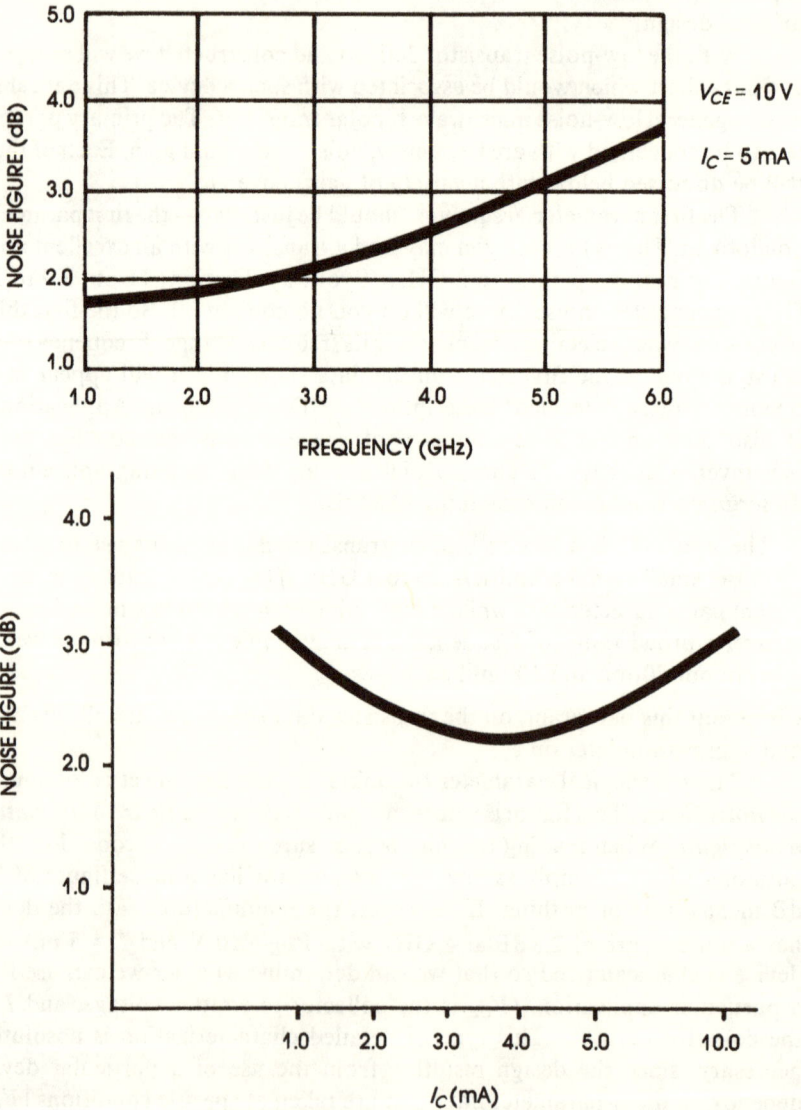

Figure 5.8 Noise Figure *versus* Frequency and Collector Current

Figure 5.8(b) is noise figure versus collector current at a fixed frequency (3.5 GHz in this case). If you compare I_c = 5 mA in Figure 5.8(b) to 3.5 GHz in Figure 5.8(a) you will see that the same noise figure results.

Some manufacturers use *noise contours* to show the noise figure of their devices. Noise contours are a graphical representation of the noise figure presented on a Smith chart. Figure 5.9 shows just such a chart with various noise figures presented on it. Notice that there are different points where a specific noise figure can be achieved. The one parameter that may suffer, however, is the input VSWR, if the input matching procedure is not treated carefully. The contours are derived by using the admittance parameter and optimum noise figures for the device and finding the circle center, the angle of the vector, and the radius of the circle. With this information, the admittance points can be picked from the chart for the noise figure desired, and the transistor can be *noise matched* for optimum noise performance. Gain will be somewhat lower than the maximum gain possible, but the noise performance will be excellent.

With the noise characterized, the next parameter to be investigated is the *transistor gain*. With this parameter we are looking at two values: G_{MAX} and G_{NF}. G_{MAX} is the gain from *conjugately* matching the device for a specific gain. The noise figure obtained with this condition depends on the characteristics of the particular device used. A conjugately matched transistor will produce the maximum gain available.

G_{NF} is the gain achieved when the transistor is matched specifically for the optimum noise figure at that particular frequency. This gain is significantly lower than the gain achieved at G_{MAX}. To illustrate, consider the comparison of G_{MAX}, G_{NF}, and noise figure for three devices in Table 5.1.

Table 5.1

Device	Frequency	NF	G_{MAX}	G_{NF}
A	4 GHz	2.8 dB	11.0 dB	7.0dB
B	4 GHz	2.7 dB	11.0 dB	8.0 dB
C	4 GHz	3.0 dB	13.0 dB	10.0 dB

These data are from three commercially available devices designed to provide basically the same performance. When designing for gain with these devices, considerably more gain is received than when designing for optimum noise figure. In most cases there is a difference of 3-4 dB in gain. This difference, of course, is a logical result, since the design is around a conjugate

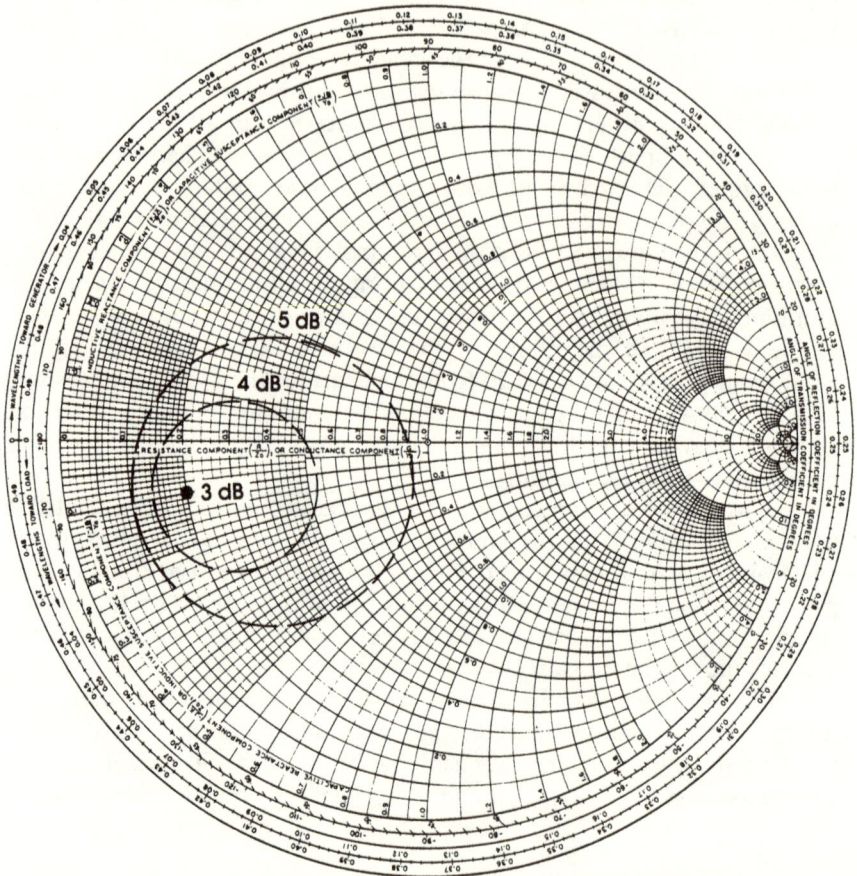

Figure 5.9 Noise Contours

match for power gain and around noise admittance parameters for a noise
match. This latter type of matching does not guarantee a maximum transfer
of power and thus will not guarantee a maximum amount of gain through
the device.

One additional gain figure is sometimes given on transistor data sheets.
This figure is the *insertion power gain*, designated as $|S_{21}|^2$. This gain is
obtained by dropping the device into a circuit with a 50-Ω line at the input
and a 50-Ω line at the output. This gain is useful for wideband operations
where a wide range of matching is not feasible using a conjugate match. The
value of $|S_{21}|^2$ for the three previous devices is 5.5 dB, 6 dB, and 7 dB,
respectively. About the same gain is received as with optimum noise figure

with $/S_{21}/^2$. However, the good noise figure will not be recieved as when performing a specific noise match.

Gain contours can be drawn just as we drew the noise contours. Figure 5.10 shows these contours for one of the devices referred to previously. When we superimpose the gain and noise contours on one another, we have a picture of the capabilities of a device and can pick our operating point to comply with our specific requirements for noise and gain. The combination of noise and gain contours is shown in Figure 5.11. One restriction should be made at this point concerning the gain contours. Note in Figure 5.10 how the circles are all inside the boundaries of the Smith chart. Within the boundaries means that the device will be unconditionally stable under these conditions. A gain contour that is not completely inside the chart tells us that this condition has the possibility of becoming unstable and going into oscillation under certain conditions. So take care when designing to be sure to be in a stable area of the chart.

Thus the low-noise bipolar transistor can be chosen by referring to a data sheet for frequency, noise figure, and gain. With these parameters chosen and matched to a particular device, you can then consider such areas as how much current the device will use, power dissipation, and a package style that will fit your particular application.

5.1.2 Linear Bipolar Transistors

When transistors are biased for a particular application they are usually biased either class A, B, or C. Class A is biased well above cutoff, so that the transistor is conducting over the entire 360° of the input cycle. Class B is biased at cutoff and the device conducts for only one half (180°) of the input cycle. Finally, class C is biased well below cutoff (usually 2½ times cutoff) and only conducts for 120° to 150° of the input cycle. Generally, class-A operation is reserved for low-power levels where noise is the primary concern; class B is used for push-pull operation of power stages; and class C is almost always used when high power is needed.

It is possible, however, to get a higher power output by biasing some transistors Class A. These are the microwave linear transistors. Devices are available that produce 7 W at 2 GHz or as much as 2 W at 3 GHz. This capability may be very impressive, or it may mean nothing without the understanding of what a linear power transistor does.

To aid in understanding the linear device, consider a low level transistor stage. It may be operating in the microwave range or in the audio range. The device is built with its associated components, proper bias is applied, and a signal is placed at the input. With these conditions satisfied, a certain

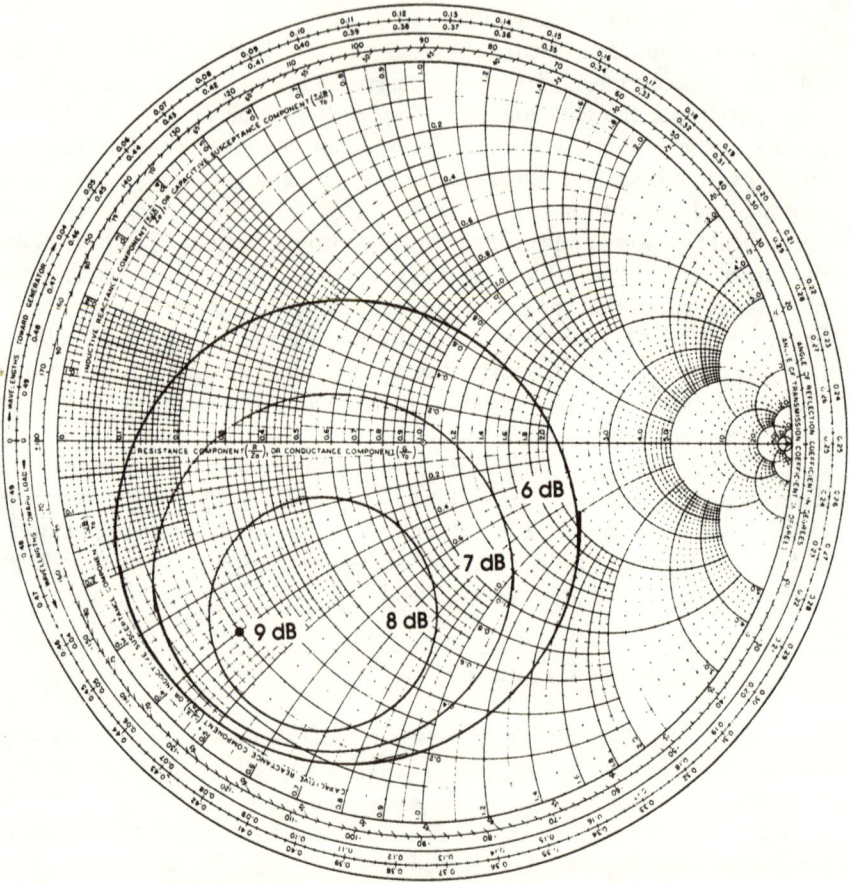

Figure 5.10 Gain Contours

output will be obtained. If the input level is increased, the output level will follow according to the characteristics of the transistor and its circuitry. This increased output will increase until a point called the *compression point* is reached. This is where the output does not increase at the same rate as it previously did. A further increase in the input level will result in less and less level at the output. This is shown in Figure 5.12. If a 10-dB increase is applied to the input and there is only a 9-dB increase at the output, the 1-dB *compression point* has been reached. This situation is one which all class-A biased circuits encounter. The compression point is what limits the amount of power available in most class-A circuits.

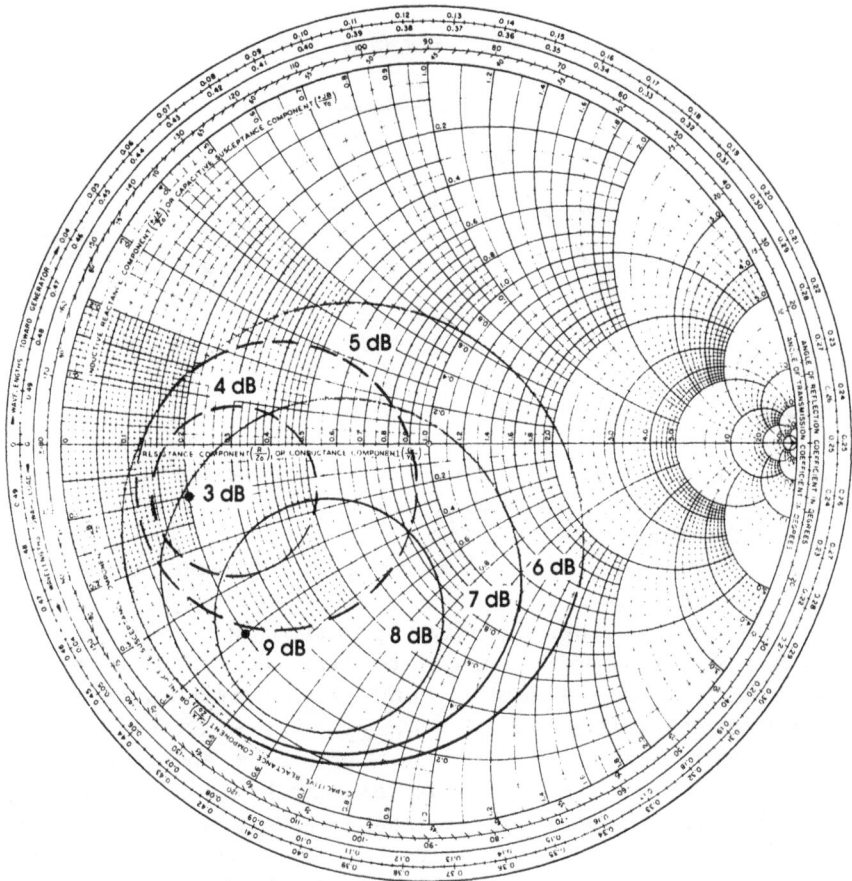

Figure 5.11 Noise and Gain Contours

In class-C operation the device is biased well below cutoff (usually 2½ times), as we previously stated. With this type of biasing, low-level inputs will have no effect on the output, since the input level is not large enough to overcome the cutoff voltage applied for class C. As the input power level is increased more and more, it will approach the cutoff point of the transistor. Once the bias has been overcome, the input level will be sufficient to turn the device on. The transistor will then snap on and display a constant power output which will increase very little, if any, as the input power is increased further. If you have an input power which is required to vary over some range and needs the power output to vary, neither the class-A device (which is

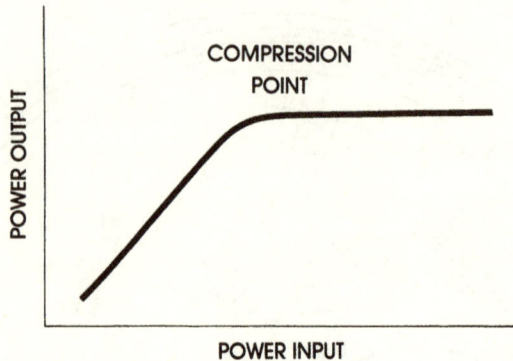

Figure 5.12 Compression Point

limited in power level) nor the class-C device (which does not vary the output power) will do the job. A different device must then step in and do this task — the linear power transistor.

The linear power transistor is the bridge needed between the high-power class-C output stage and the low-level class-A input stage. It enables a designer to use class-A operation (linear operation) right up to the final power stages. It should not be assumed that the linear power transistor competes with the class-C power transistor. On the contrary, the linear device compliments the power stage by providing very good drivers for the high-power stage. The linear-power stages are unique in that they are able to operate class A, but still are built rugged enough to survive the high collector currents needed for power amplification. Methods used to accomplish this are, of course, proprietary and thus cannot be discussed at this point. Suffice it to say that the transistors are a unique blend of low-level operational characteristics combined with the stout construction needed for higher power.

Items from a typical linear-power transistor data sheet are shown below:

Frequency	1 - 2 GHz
Power Output	1.6 W @ 2 GHz
VSWR Mismatch	∞
Linearity	–0.2 dB, +1.0 dB

All of the parameters listed above have specific conditions associated with them, just as the low noise transistor did. An example of these conditions may be V_{CE} = 20 V, I_C = 200 mA, and frequency = 2 GHz. These conditions indicate how the data you are viewing was taken and what it takes to duplicate it. A brief explanation of each of these parameters is shown below.

The first parameter, *frequency*, gives the range over which the transistor will operate. Frequency does not indicate, however, that the device will cover this entire band in one shot and have a response that will be flat within 0.5 dB. It only says that the frequency range given is the one the transistor was designed to handle. Some devices will be characterized as operating from 1 to 2 GHz, for example, but will not have a constant output power over this band. They will only have the required output over a narrow band — 1.0 - 1.15 Hz, 1.6 - 1.7 GHz, etc. The power output versus frequency curve in Figure 5.13 is a typical example. It shows that approximately 1.9 W is available from 1 to 2 GHz. In reality you will not be able to match the device over this broad a band. You will have to settle for something less. So check the data sheets carefully to see just what frequency is being referred to.

Power output is the next important parameter to be checked when choosing a linear transistor. As we have said previously, certain conditions are spelled out when specifying parameters. A device cannot be said to deliver 1.7 W of power at 2 GHz without defining the conditions under which the 1.7 W was obtained. What is needed is such data as V_{CE}, I_C, and power input. You will recall that these parameters were shown in Figure 5.13 where the power output *versus* frequency curve was shown. The power on that graph was only valid for those conditions.

Another curve which illustrates how important these parameters are is shown in Figure 5.14. If we take the case of V_{CE} set at 15 V, we can get anywhere from 0.5 W to approximately 1.6 W by varying the emitter current (I_E) from 100 to 300 mA. If we increase V_{CE} to 20 V, the same range of current yields from 0.6 W to about 2.25 W. So you can see how important it is to specify the operating voltage and current to make the parameter repeatable. Therefore, the power output on the data sheet should always include test conditions, including frequency.

VSWR mismatch: One of the most important properties of a device handling power is its ability to perform under mismatch conditions. If, for example, a device is producing 2 W (or a large majority of it), then this power can be reflected back to the device. If the device is not equipped to handle this reflected power, it will be destroyed. This will result in lost time waiting for a replacement device plus the cost of a new device, neither of which are conducive to schedule and budget constraints. So always check to see what sort of a mismatch tolerance the device exhibits. The example we have

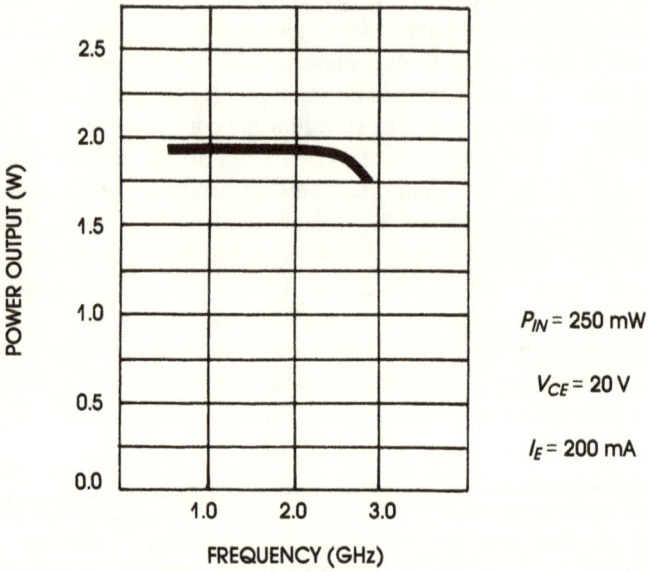

Figure 5.13 Power Out *versus* Frequency

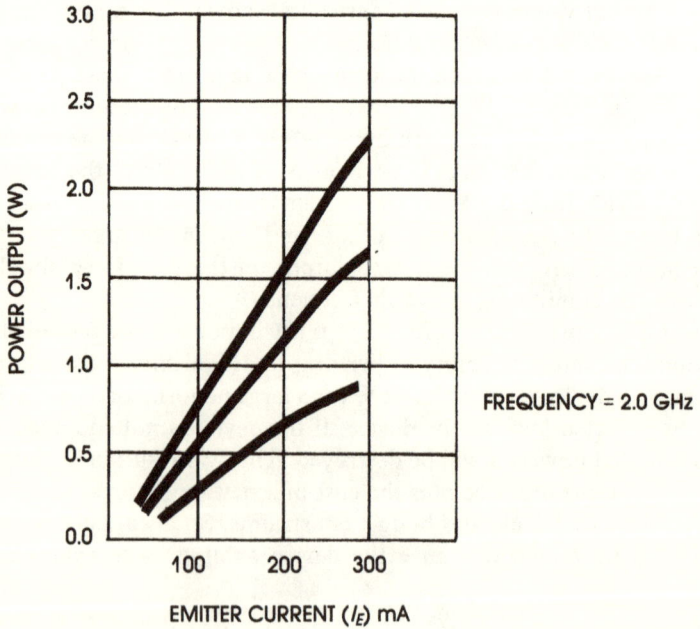

Figure 5.14 Power Out *versus* Current

presented shows infinity as the specification for mismatch, meaning that a VSWR of infinity (or an open circuit) will not affect or damage the transistor. If there is any chance of a large mismatch present at the output of a power stage, specify a transistor which will take an infinite mismatch.

The final parameter on our typical data sheet is *linearity*. Linearity is a natural parameter to investigate since we are talking about linear devices. Linearity of a device is how close its output response tracks a straight line drawn between a low- and high-power end of its operating range. The data sheet device showed –0.2 to +1.0 dB of variation over this range. The conditions given for this parameter are usually frequency, power #1 (low end) and power #2 (high end). Typical numbers might be 2 GHz, P_1 = 0.15 W, P_2 = 1.5 W. If the response of P_{IN} *versus* P_{OUT} is plotted, and a straight line is drawn between P_1 and P_2, the variation from this line will be the linearity. The ideal figure, of course, would be ± 0 dB.

The microwave linear-power transistor, then, is the device that produces an intermediate level of power amplification while operating in a linear, or class-A mode. Whenever a range of linear-output power levels is needed, this transistor is the type to use. After investigating the parameters above for your particular application, choose the linear-power device which will perform up to these parameters.

5.1.3 Power Transistors

The microwave power transistor is a prime example of good things coming in small packages. Here is a device which has dimensions less than one-half inch square which will deliver upwards to 25-30 W of power CW at frequencies up to 2 GHz. This truly is a marvel to behold. To aid in your understanding of these tiny powerful devices, we will now proceed to build a bipolar power transistor from its silicon wafer first step to the protective cap which is epoxied to the case to protect the internal circuitry.

As mentioned, the power transistor begins with a silicon wafer. The silicon provides a medium for the transistor with certain properties that allow current to flow. These properties can be compared to a vacuum in a vacuum tube. For the tube, the vacuum is the medium; for the transistor, silicon is the medium. The wafers used are epitaxial. Remember from our discussion of low-noise devices that epitaxial is the "growing" of a substance on the same type of substance; that is, the growing of a layer of silicon substrate. The silicon is grown in ingots 1 to 3 inches in diameter and from 6 inches to several feet in length. A single crystal "seed," ¼-inch square by several inches long, is used to start the growth. The seed is rotated and slowly withdrawn from a crucible of molten silicon that is doped to be *n*-type or p-type material by the addition of basic elements. Ingots of n-type doping use antimony or arsenic.

The ingots are grown at temperatures approaching 1500°C. The dopants must all be distributed uniformly to prevent excess concentrations in one area and deficiencies in others. The ingots are sliced with a diamond saw into wafers to a thickness of around 0.015 inch. They, then, are mechanically and chemically polished to a mirror finish that reduces their thickness to around 0.012 inch.

The polished wafers are then placed in an epitaxial reactor at about 1200°C, and a doped flim of silicon is grown on top of them by means of a chemical vapor deposition. The doped film may be as thin as 0.0001 inch (0.00025 cm) or as thick as 0.002 inch (0.005 cm). This film can be either n or p type, depending on the type of device being fabricated. The resulting material is a *silicon epitaxial wafer*, which is probably the most important section of the transistor, since everything is constructed on top of this wafer. It is somewhat like the foundation for a building. The foundation must be right, or the entire building will crumble. Similarly, if the epitaxial wafer is not grown properly, the transistor will not operate properly. The wafer is now complete, and the next step in our building process includes *oxidation* and *masking*.

The microwave power transistor requires a large number of narrow emitter images to obtain a high emitter perimeter to base-area ratio because a large current is drawn by power transistors, and there must be provisions within the device to handle this current. An absolute minimum practical width for the emitters is one micron (0.0001 cm, 0.000039 inch). The base, the emitter, and their contact patterns are defined in an oxidated silicon wafer by a photoresist and etch process (print and etch). The 1-μm geometries necessary for device operation must be maintained by this photoresist and etch process. The photoresists, which are polymers sensitive to exposure to ultraviolet light, are used to define the necessary pattern. These patterns may be either positive or negative.

When positive patterns are used, the exposed area is developed, and the resulting image is the same as that which was exposed. Thus, if a hole in the oxide to permit diffusion of the impurities is required, the mask would have clear areas with the pattern that is required. The remainder of the mask would be a dark field. This arrangement was once a very difficult item to align because the operator could not see through the dark field to align the mask. Now, however, there are see-through masks that are opaque to ultraviolet light, which makes the positive masks much more practical and useful.

Negative resists reach, of course, in the opposite way of the positive mask. That is, the unexposed area is developed, rather than the exposed area.

Organic solvents are used to develop these resists that cause them to swell. When this swelling occurs, the openings shrink and all edge definition is lost. Various types of rinses can be used to attempt to reduce this swelling, but as the resist shrinks, its adhesion to the substrate is degraded and some may even peel off. Therefore it is desirable to use the positive resist when small geometries are being used.

Thus the positive resist has the following advantages for the small geometries used in power transistors.

- It does not cause swelling, so a much thicker resist can be used for better definition.
- The processing is easier, since the positive resist does not have to be postbaked at high temperatures to ensure adhesions.

The oxidation process is performed to protect the surface of the device. All diffusion takes place through "windows" cut in the oxide to allow the *n*-type dopants to diffuse into the silicon. As previously described, photosensitive material is applied to the top of the wafer. The photoresist is exposed when a light source is shown on the wafer, similar to the everyday process of developing a photograph from an instant camera. The mask (positive) shields the window from exposure. When developed, the unexposed window is removed, since light is required to harden the photoresist.

Exposing the silicon for dopants to be diffused requires an etchant that attacks only the SiO_2 (silicon dioxide) and not the silicon or the hardened photoresist. That etchant is hydrofluoric acid (HF). Results of such an etchant and the oxidation process are shown in Figure 5.15. This figure also shows the next phase of our power transistor construction — base and emitter diffusion.

Diffusion is officially defined as the movement of carriers from a region of high concentration to regions of low concentration and was described previously in our section on junctions. Diffusion can be likened to dropping a stone in the water and seeing the ripples move from the point of impact out until the water is smooth again. A high concentration (high amplitude) of water is in the center with a movement outward to the end of the large circle where a low concentration (low amplitude) is, which is the same general idea of diffusion in transistors. In the case of transistors where silicon and silicon dioxide (SiO_2) are concerned, doping impurities diffuse into the silicon at elevated temperatures to form the desired junctions. The same impurities penetrate the silicon dioxide much more slowly, and, therefore, silicon dioxide on the surface of the silicon (shown in Figure 5.15) acts as mask to determine the areas into which diffusion occurs.

As seen in Figure 5.15, base diffusion occurs before emitter diffusion. The base of an *npn* device is *boron*, and the emitter is *phosphorus*. Diffusion

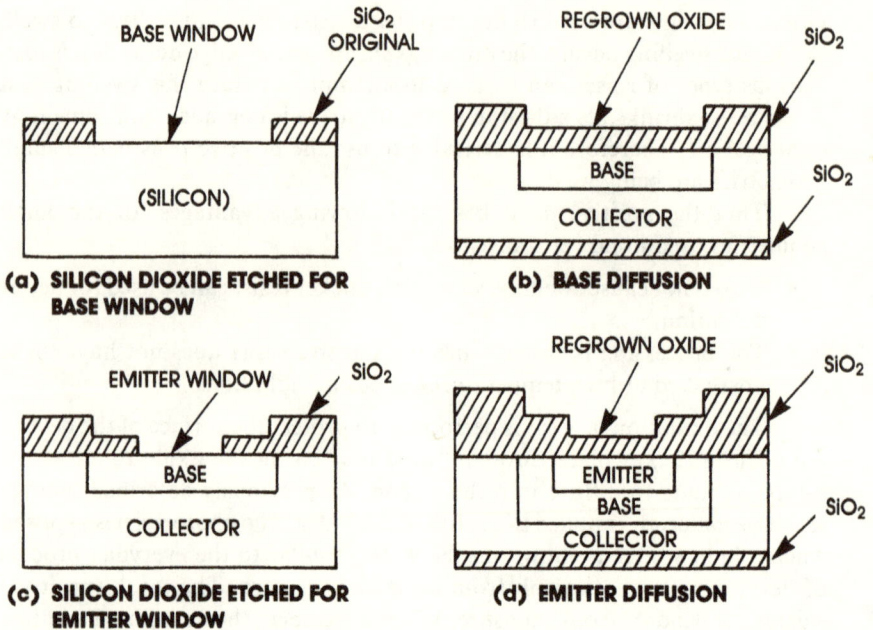

Figure 5.15 Oxidation

temperatures are high, as referred to earlier, and are on the order of 1000°C. This high temperature causes the dopant (either boron or phosphorus) to diffuse from areas of high concentration to those of low, which satisfies the definition. Diffusion may be accelerated either by high temperatures or higher concentrations. As might be expected, the farther the dopant travels into the transistor, the lower the concentration. Whenever a high dopant concentration region exists (on the surface, for example) a low resistance exists, which results in a high concentration of carriers (electrons or holes).

The diffusion process, and how it occurs, determines the quality of the finished transistor. Once the base is diffused, the emitter-base junction must overtake the collector-base junction. That is, the base thickness must be controlled by controlling the separation of the emitter-base and collector-base junctions. This thickness will greatly affect the parameters of the transistor. By controlling the base thickness and the junction-areas dimensions, the f_{MAX} (maximum frequency of oscillation) is improved greatly. Maximum frequency of oscillation is a measure of the ultimate frequency capability of the device; that is, how high in frequency the device will work. Obviously, when discussing microwave devices, this frequency will have to be as high as your application requires. In any case, it must be well within the microwave spectrum to be useful. The goal of this critical base region is to keep it thin. This thickness and the emitter depth can be controlled best when the doping densities are large; that is, the area of low resistivity where concentrations are high. Excellent operation occurs when the emitter depth is ½ to ¾ of that of the base.

The final process before we put the device in a package and attach leads to it is that of *metallization*. There are two schools of thought on metallization. One says that aluminum is the best metal for microwave power transistors, the other says gold provides superior performances. There are advantages and disadvantages to both materials, and for that reason, both will be covered. The first to be covered will be aluminum. (This order does not mean that aluminum is any better or worse than gold. The only criteria used to make this decision was that "A" comes before "G" in the alphabet.)

The aluminum metallization technique is illustrated in Figure 5.16. Note that the top portion of the figure is similar to that at the bottom of Figure 5.15 after we finished our diffusion and deposition processes. Aluminum is evaporated on the wafer to something less than 2%. An electron-beam technique is used to achieve a uniform and pure aluminum layer. A photolithography process, described earlier, is used to define the final metal pattern needed. A patented etch process is used to etch away the unwanted aluminum. This technique is very powerful in that it allows extremely fine aluminum lines of less than 0.1 mil to be cut. In addition, the process is inherently very high yield and defect free. Its major advantages are finer line geometries and very high yield and defect free. Its major advantages are finer line geometries and very high yield on large devices (permitting very high-power, high-frequency units).

Aluminum-silicon metallization is another process employed in which there is an addition of from 1 to 4% silicon to the aluminum. The advantages are the same as with the aluminum system, but there is the additional benefit of vastly improved yield and reliability on very high-frequency devices. Usually when a transistor is said to have aluminum metallization, it, in fact, has aluminum-silicon metallization.

Gold, it is said, offers distinct advantages over any aluminum system. It is less prone to electromigration (15 to 20 times better), it eliminates the corrosion problems encountered with aluminum, it can be etched easily (it does not undercut), and it virtually eliminates the microcracking associated with relatively brittle aluminum.

All of these advantages mentioned are applicable to pure gold — but pure gold cannot be deposited on bare silicon. Silicon and gold form a eutectic around 400° C, a temperature not compatible with transistor processing and assembly. Thus a problem arises because the gold migrates into the silicon and no clear boundary is formed between the two.

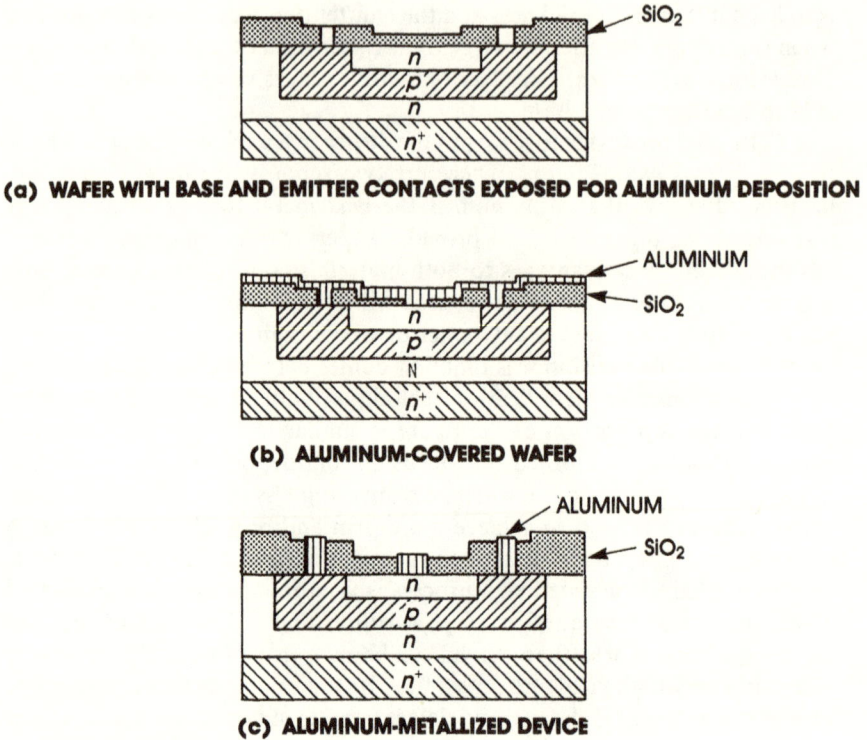

(a) WAFER WITH BASE AND EMITTER CONTACTS EXPOSED FOR ALUMINUM DEPOSITION

(b) ALUMINUM-COVERED WAFER

(c) ALUMINUM-METALLIZED DEVICE

Figure 5.16 Aluminum Metallized Device

A high-temperature metal, therefore, must be used with gold to act as a barrier separating the gold from the silicon. This barrier metal should be a good barrier for gold (gold should not diffuse through it); it should make good, uniform electrical contact to the silicon; and it should adhere to the oxide of the device to allow for subsequent assembly.

In actual practice, no single element satisfies all of these conditions, so a multimetal system must be used. Some of the systems most commonly used as multimetal barriers follow.

Platinum Silicide-Titanium-Platinum-Gold — Platinum silicide is used to provide good electrical contact, titanium is used for adherence to the oxide, platinum is used to provide a barrier to gold diffusion, and gold is used to carry high current densities. This system is primarily used in beam-lead applications. It is very complex and impractical for very small geometries because of the etching difficulties of both titanium and platinum. This system, if it does work, is very reliable, if care is taken to passivate the titanium, which is worse than aluminum in terms of corrosion.

Platinum Silicide-Molybdenum-Gold — This metallization system has been used for several years in small-signal high-frequency transistors. The main problem in this system is the high electrical contact resistance. In high-power transistors, low input resistance is one of the essential features for achieving good performance. Platinum silicide is used to minimize this problem. Getting consistently low contact resistance, however, still remains a problem to be solved.

Platinum Silicide-Tungsten-Gold — This system has been introduced for use with high-power high-frequency transistors. Tungsten does not make good electrical contact to silicon, so platinum silicide has to be used to lower the contact resistance. Tungsten also does not adhere well to oxide. To overcome this problem, a very high sputtering energy, which tends to degrade the E-B junctions and, thereby, results in very poor yields, is required. To solve this problem, a fourth metal is required to "glue" the tungsten to the oxide. Tungsten, although better than titanium, is not an acceptable barrier to gold, as accelerated tests at 600°C have proven. Moreover, tungsten is very hard to etch, as it undercuts the gold during etching, but it is impossible to inspect the quality and the seriousness of the damage. How far it has been undercut is clear only when the gold falls off the wafer. Obviously, this type of metallic barrier is not used very often.

An additional composition of platinum silicide-titanium-tungsten-gold has also been shown to be very successful. The platinum silicide is used to provide a stable ohmic contact to silicon. The tungsten layer is also used as a barrier against gold diffusion into and eutectic alloying with silicon, as well as providing adhesion of the gold to the silicon and silicon dioxide (SiO_2).

As we stated earlier, each of the metallization processes has advantages and disadvantages. Generally selection is a matter of preference and application.

We now have a power transistor chip constructed and ready to drop into a package. The package is probably every bit as important as the chip we have just built. It does much more than simply giving a means of attaching the transistor to a circuit. Three factors must be taken into consideration:

- Lead inductance must be kept to a minimum. Concentration must be on the common lead first. This may be the emitter or, more commonly, the base in power devices. Second priority for reduction in lead inductance is the input and last is the output.

- Thermal dissipation of the case must be high. The most important criterion when operating power devices is to get the heat away from the chip and out to the chassis. This dissipation occurs when the package does its job.
- The package should be a reliable and easy-to-use device. Such factors as package type, shape, size, and material should all be considered.

As the frequency of operation of any component increases, you must become increasingly aware of the inductance involved with that component. The need for awareness becomes evident when examining the equation for inductive reactance ($X_L = 2\pi f L$). If this reactance is to remain low as the frequency increases, the inductance must be decreased or kept at a minimum. In an actual device every bonding wire from the chip is the same size and, more importantly, the same shape. This ensures good consistency within the device, low loss through the wire, and low inductance.

As we have stated previously, the most important criterion when operating power devices is to get the heat away from the chip and out to the chassis. The thermal characteristics of the chip are largely dependent on the particular design. From heat transfer theory, a point heat source (the chip) dissipates heat through a solid in the shape of a cone. That is, the heat is concentrated at the origin and fans out as it travels from that point. This definition is similar to our definition of diffusion with a high concentration moving to a lower concentration. Another way of looking at heat transfer is to consider a beam from a flashlight and how the light fans out and diminishes as you get farther away from the bulb (the source).

By understanding how the heat dissipates through a substance, you can see that if the chips (or sources) are too close together, there will be overlapping heat areas that produce "hot spots." These hot spots are areas of high heat density that make it very difficult to get heat away from the chips at a sufficient rate to protect them.

To eliminate these hot spots, or interference areas, the cells of a chip can be spaced such that these areas exist in low-heat density areas. A high-yield construction is used in this case, and a geometric symmetry arises from the use of cellular and interval matching techniques, virtually assuring thermal balance. As an example, thermal measurements were taken in a 32-cell, 125-mil × 75-mil chip and showed thermal gradients (termperature differences) of only 3% from cell to cell. Single-cell transistors often show temperature variations of 50°C from center to edge. Uniform heat distributions allow the junctions to operate at moderate temperatures, assuring long life reliability of the device.

We have now come to the point where the transistor is in its package. This package may be one of a variety available from different manufacturers. The method of mounting the transistor should be carefully considered. Whether it is a flange-mounted device or studded device, there are specific ways of properly mounting each. The following general rules should be followed for all devices. For specific requirements, consult the vendor data sheet.

- Be sure the metal area that the device is to be mounted to is flat and smooth.
- Remove all burrs from any mounting holes.
- Be sure the metal areas are clean — remove all oil and chips.
- Use thermal grease on the heat-sink area.
- Align the transistor properly; that is, be sure which lead is the base, emitter, and collector.
- Using the proper-sized screws for the flange-mount device and appropriate washers, mount the device to the heat sink.
- If a studded device is used, be sure to use the right torque when putting the nut on the stud. Too much torque will break the stud from the transistor. The following torques are recommended for various studs.

Table 5.2

Stud	Torque
¼ inch	5 ±1 in · lb
⅜ inch	8 ±1 in · lb
½ inch	10 ±1 in · lb

By using the proper mounting techniques, many potential problems will be eliminated both on initial turn on and further down the road when the heat buildup destroys areas of the device.

To finish up the topic of power transistors, we will look at a typical data sheet. Naturally, the first item to look at would be **output power**. This value is usually given as a power at a frequency; that is, 20 W at 2 GHz, for example. For this reason it is advisable to consult other pages of the data sheet to see what the device will do at other frequencies. Figure 5.17 shows a power out *versus* frequency curve for a 20-W device. Notice the difference in the output levels with various input power levels. Obviously, this device was designed to operate at a 20-W level since that is the only place where the response is flat. All other input levels result in a lower response at the output.

The second area to look at when deciding which device to use is the gain of the device. The device may not be much good if, for the 20-W output, 15 W must be the input. Some reasonable amount of gain must result from the device if it is going to be worthwhile. A gain figure of 6 dB is a typical number for a device such as the one we have been discussing (20 W at 2 GHz). This gain will mean an input of 5 W to get our 20 W out. These numbers correspond to the numbers in Figure 5.17.

Another area to be considered is the efficiency of the device. *Efficiency* is defined as the ratio of power out to power in. The power out is our 20 W. The power in is our 5-W input plus the dc power needed to operate the transistor. Typical efficiencies will run from 30 to 40%, which means that we must put in approximately 50 W to get 20 W out. So, 5 W is input power, and 45 W is dc power. If we use a 28-V supply, we are allowed to draw 1.6 A of current (P = EI, 45 = 28 V and 1.6 A). These numbers will give the device an efficiency of 40%. Look very closely at efficiency numbers, because as much power out as possible with an absolute minimum of power in is wanted.

A final item to consider when choosing a power transistor is the input and output impedances of the device. Figure 5.18 shows the impedances for a 20-W device, and Figure 5.19 shows them for a 2-W device. It should be pointed out that the 2-W device is not a linear power transistor but is a class C power transistor. The two are shown to illustrate how the impedances can vary with the power level. One item to note is how low the impedances are that have to be matched to a 50-Ω source or load. This is why a good-sized microstrip pad is usually seen at the input and output of most of the higher power circuits to accomplish this match.

So, in review, consider the power output of a device, its gain, its efficiency, its input and output impedances, and make provisions to get the heat out of the device and into the chassis. If all of these items are taken into consideration, a reliable operating power amplifier that will perform as expected will result.

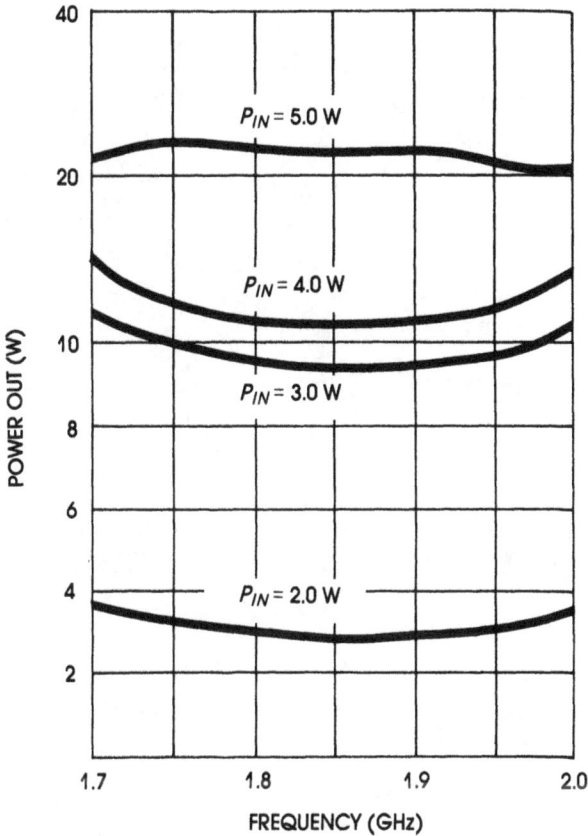

Figure 5.17 Power Out *versus* Frequency

Thus, the bipolar transistor for microwave applications has been discussed in three distinct and individual areas: low-noise, linear power, and power. Each has specific areas where it can be used and has found areas where only microwave tubes previously were used. Truly the bipolar microwave transistor has a prominent place in microwaves today.

5.2 FIELD EFFECT TRANSISTORS

The overwhelming choice for microwave transistors in the mid-1970s was the gallium arsenide field effect transistor (GaAs FET). You could not pick up a microwave magazine without seeing more and more advances being made in these devices. Their popularity and usage grew and their acceptance

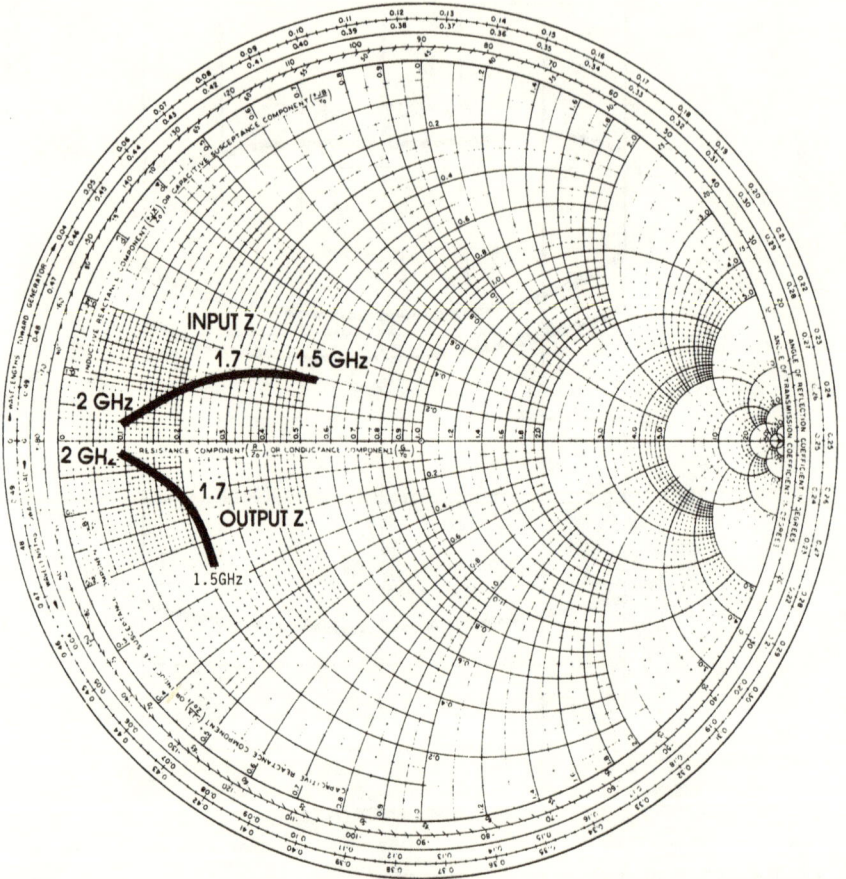

Figure 5.18 Impedances for a 20-W Device

increased with good reason. They were, and are, excellent devices for microwave applications.

The field effect transistor differs from the bipolar in more than name. You will recall from our discussion of bipolar transistors that the device operated with two types of carriers, electrons and holes. The field effect transistor, on the other hand, is what is termed a *unipolar* device, that is, it operates with only one type of carrier — electrons. As discussed in Chapter 3 on junctions, this reliance on only one type of carrier allows the devices to operate at significantly higher frequencies than their bipolar counterparts. This, of course, makes them an ideal choice for microwave applications.

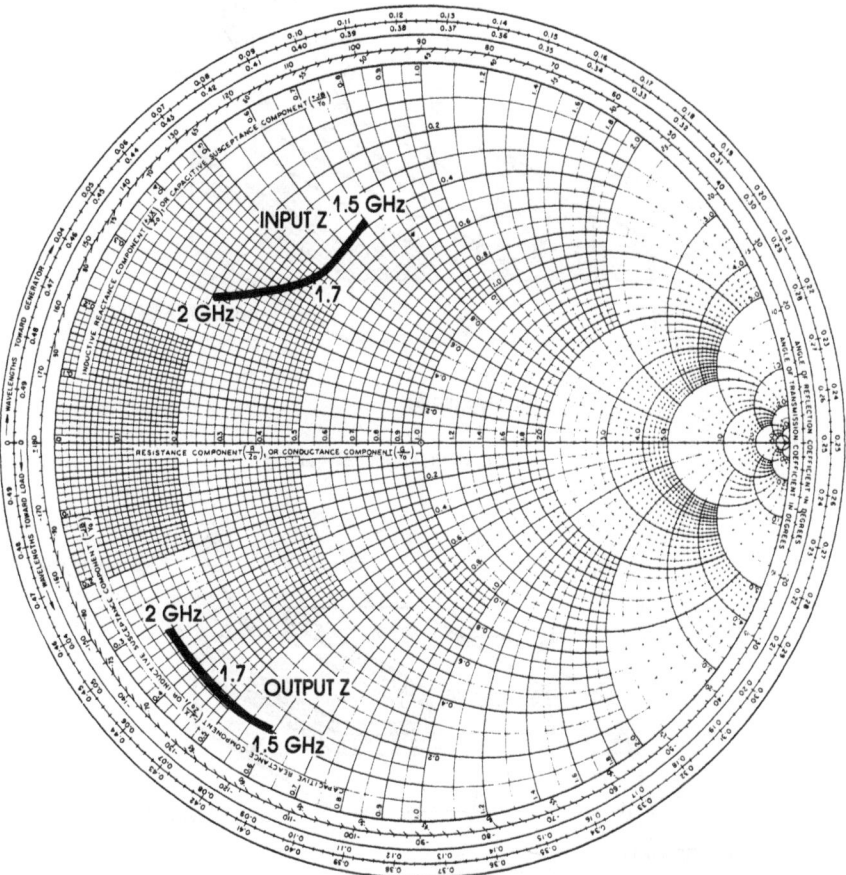

Figure 5.19 Impedances for a 2-W Device

We have stated that the mid-1970s saw the blossoming of the FET on the microwave horizon. These devices actually have a much longer history than that. The theory for FETs goes back to 1926, and William Shockley first proposed them in 1952. At that time, however, there were many technological and fabrication difficulties that kept the FET from arriving on the scene until the early 1960s. At this time the silicon bipolar device was well defined and pushed the FET further out to the 1970s.

The earliest FET was the junction FET (JFET), which became available about the same time as the first microwave bipolar transistors. The JFET construction is shown in Figure 5.20(a). Advances in techniques and a need

SOURCE GATE DRAIN

n-TYPE MATERIAL

p-TYPE MATERIAL

(a) JUNCTION FET (JFET)

GATE

SOURCE DRAIN METALLIZATION

p-TYPE MATERIAL

n-TYPE MATERIAL OXIDE

(b) METAL-OXIDE-SEMICONDUCTOR FET (MOSFET)

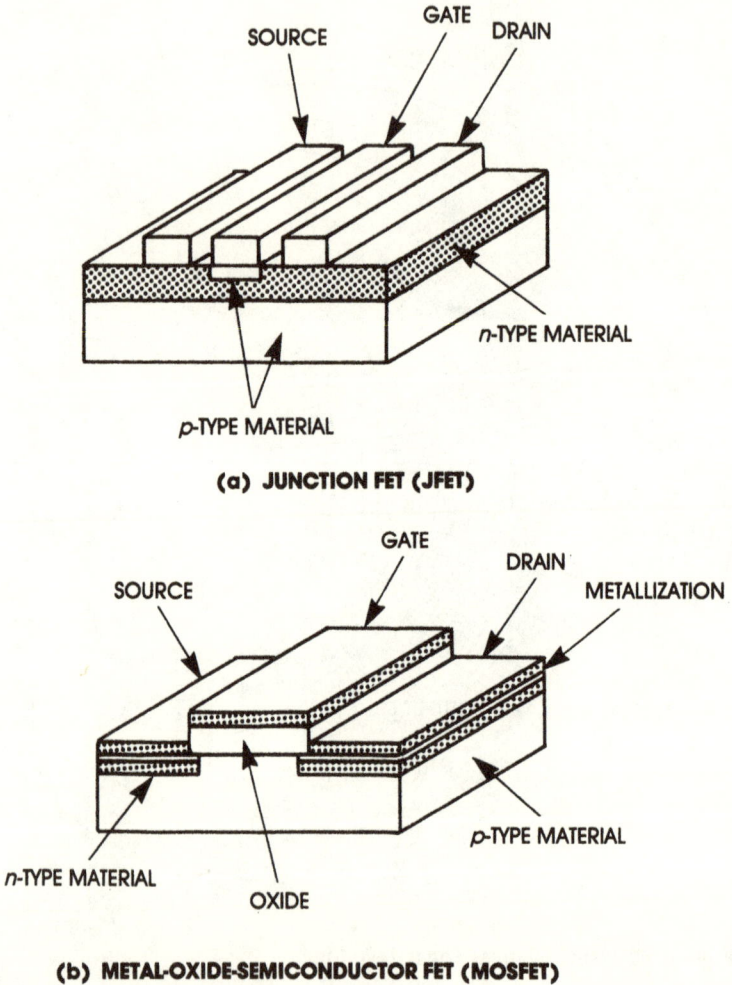

Figure 5.20 Field Effect Transistor (FET)

for low power resulted in the metal-oxide-semiconductor FET (MOSFET). This device also may be referred to as an insulated-gate FET (IGFET) because of the oxide layer between the gate and the substrate as shown in Figure 5.20(b).

These FETs offered no competition to the well-established bipolar transistor in either of the forms shown in Figure 5.20. When the metal-semiconductor FET (MESFET) came on the scene, however, the bipolar began to notice some competition. This device now introduced the Schottky barrier as the gate of the device and completely departed from the conventional PN junction. Figure 5.21 shows a representation of the MESFET. With the construction came the now familiar gallium arsenide field effect transistor (GaAs FET). This was the device which was the buzz word of the 1970s and on into the 1980s.

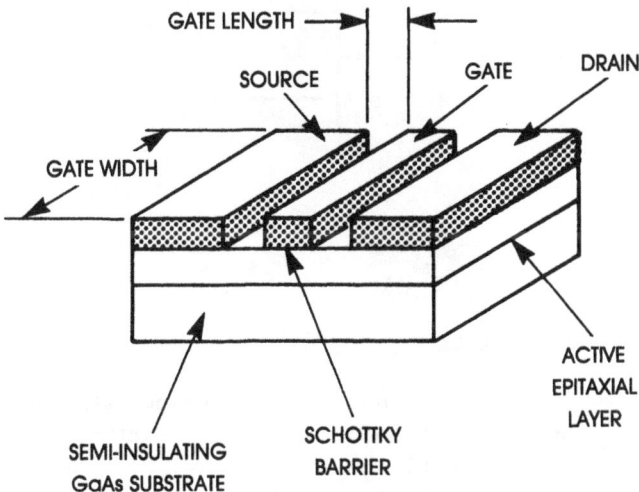

Figure 5.21 MESFET

Designation of the elements are different from those of a bipolar device. When using a bipolar device you put the 10 V or 12 V on the collector, biased the base, and attached the emitter to ground. In a FET you put the 10 or 12 V on the drain, bias the *gate*, and attach the source to ground. This is illustrated in Figure 5.22.

Presently GaAs FETs are in use in many areas of microwaves. Their range of frequency is upwards to 40 GHz and above in many cases. Noise figures of 3 dB are possible in this frequency range also.

As was discussed in Chapter 2, gallium arsenide is a semiconductor compound (number III-IV) which has many advantages over the conventional material, silicon. Some of these advantages are:

1. The drift mobility is higher and, therefore, unwanted resistances which reduce gain and increase noise are lower.
2. Peak velocity of electrons is higher — thus, the high frequency gain is increased.
3. It can be made in an insulating form which allows the construction of low capacitance, high gain devices on it.

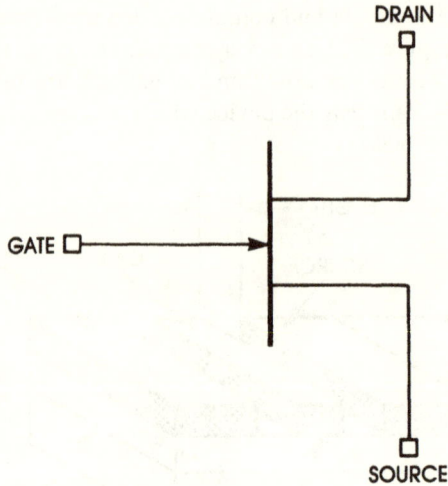

Figure 5.22 GaAs Schematic

With these advantages presented, we will now proceed to build a GaAs FET just as we did for the bipolar device. The first layer of such a device is the semi-insulating GaAs substrate which forms the foundation for the transistor. This layer consists of a doping of pure GaAs with chromium. This reduces input and output parasitics within the transistor. This can be seen as item 1 in Figure 5.23.

Diffusion of impurities from the semi-insulating substrate during epitaxial growth can degrade the electrical performance of the conductive layer that supports the transistor element (source, gate, and drain). For this reason a very thin epitaxial film is grown to prevent such diffusion. Precise control of growth rate and doping concentration is critical to maintain a uniform thickness of the layer. By controlling the diffusion of impurities, the buffer layer shown as item 2 in Figure 5.23, reduces the transistor noise figure and prevents carrier trapping. This phenomenon will cause long- and short-term failures such as I_{DSS} instability. This instability is a fluctuation in drain current (I_{DSS}), which causes the entire device to be unstable. Other terms for this action are I_{DSS} looping or stepping.

The layer which is placed on top of the buffer layer is termed *active epitaxial layer*. It is a conducting layer and may be formed in one of several ways:

- *Vapor-phase epitaxy* (VPE) — The conducting layer grown on the surface of the wafer uses gases that contain the required elements and compounds.
- *Liquid-phase epitaxy* (LPE) — The conducting layer is grown on the surface from a liquid melt of GaAs and other elements.

- *Implantation* — The elements necessary to make the top layer conductive are "shot" (implanted) into the GaAs surface using a strong electric field (50,000 to 400,000 V).
- *Molecular-beam eiptaxy* (MBE) — An insulating wafer of GaAs is placed under very high vacuum near to small furnaces called "effusers." The effusers melt gallium, arsenic, and other elements that condense on the surface of the wafer to form the active layer.

The majority of the manufacturers of GaAs FETs use VPE (vapor-phase epitaxy), although some are now investigating the use of ion implantation.

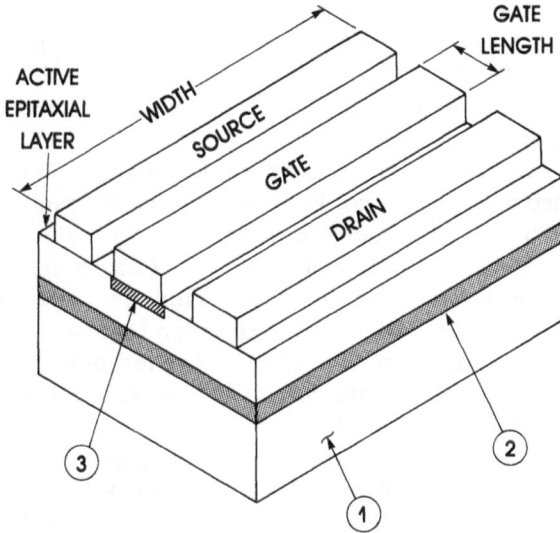

Figure 5.23 GaAs FET Construction

The source and drain of the transistor are produced by the same process and are treated the same in our discussion. These two elements are connected to each other by a conducting channel. The gate of the device modulates this channel thickness to produce the resulting power gains. The source and drain are deposited on the GaAs layer in a vacuum and then defined by a standard photolithographic process. The elements are made of alloys of gold and germanium (usually gold-indium-germanium) and are heated in a sintering process to produce low-resistance ohmic contacts.

The gate also uses standard photolithographic techniques to define its area. It is made of a layer of aluminum that forms a Schottky-barrier diode with the GaAs layer. This is shown as item 3 in Figure 5.23 and is the mechanism which allows the FET to function at the frequencies it does. The

Schottky junction was covered in detail in Chapter 3. Aluminum is used for this barrier because of its low shunt resistance, purity, ease of deposition, and the elimination of the high-temperature gate diffusion process necessary with other materials. Other metals that could be used are tungsten, platinum, and molybdenum for forming the Schottky junction. These materials, however, require a layer of titanium put on the active layer first to act as a "glue."

The metallization techniques used for the aluminum take advantage of the good characteristics listed, but disallow the formation of any metallic compounds such as "purple" plaque. This plaque can result from bonding gold wires to aluminum pads. (Purple plaque is defined as a compound that forms as a result of contact between gold and aluminum. It causes serious degradation of the reliability of semiconductor devices.) Thus, gold and aluminum should not come in contact with one another, if reliable connections are to be obtained.

Figure 5.24 is a diagram of GaAs FET geometrics. Figure 5.24(a) is a low-noise device, while Figure 5.24(b) is a power device. The source, gate, and drain of each device is masked. Notice the difference in appearance of the two geometries shown. First, both are oriented the same. That is, the gate is always between the source and the drain. This allows the gate to perform its modulating function and produce gain through the device. Second, notice the different types of structures for power and low noise. The power device has many areas for the element so that each may share the current and, thus, carry more total current. The low-noise device must be a structure which carries very little current and also has small gate resistance to keep the internal noise level low.

A point should be brought up before proceeding on with further information. That point is the importance of the gate when specifying GaAs FETs. You will notice that with different manufacturers, GaAs FETS may be called 0.5 μm, or 1-μm, or some other value devices. This number refers to the gate length and is the most important single FET element that determines high-frequency gain. Commercial, high-performance FETs have the 0.5 μm gate length and will operate to around 18 GHz. Lower-frequency FETs have 1.0 μm gate lengths and will operate in the range of 8 to 10 GHz. Some FETs have been built that have 0.2 μm lengths and will operate to somewhere around 30 GHz. Gate length, then, is the term which determines the frequency of a GaAs FET. By referring back to Figure 5.23 you can see the location of the gate length of a device.

A characteristic of the gate that comes into play when discussing power FETs is the width. Small-signal FETs may have a gate width of 0.15 mm, while a power device might have a width of 26 mm. This is also shown in Figure 5.23. The differences in width can be distinguished in Figure 5.24, where both the low-noise and power devices are shown.

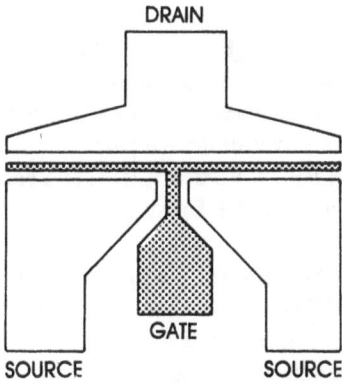

(a) LOW-NOISE GaAs FET GEOMETRY

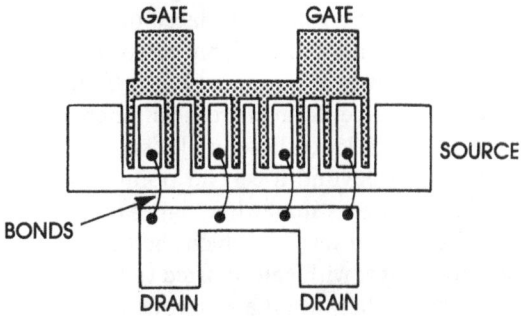

(b) POWER GaAs FET GEOMETRY

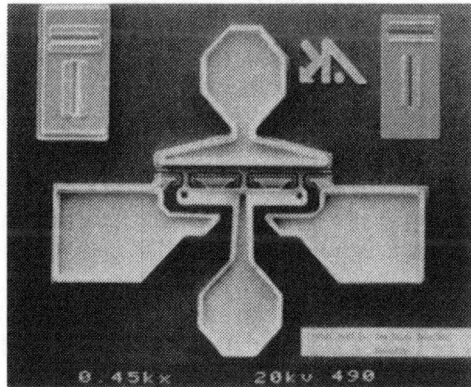

Figure 5.24 GaAs Geometry (Photo Courtesy of Avantek, Inc.)

To wrap up our building process for FETs, let us consider the package. Obviously many problems would be eliminated if a FET chip could be used in your circuit. This use, however, is not always possible. For a variety of reasons, a packaged device must be used. The package, therefore, must be such that it will have a minimum effect on the overall device performance. Figure 5.25 shows some cases used for GaAs FETs. Figure 5.25(a) is a case used for low-noise devices. A typical case for a power FET is shown in Figure 5.25(b).

Figure 5.25(c) shows a chip carrier. It is a case of sorts with a grounded source and 50-Ω input and output lines. This carrier has low thermal resistance and offers chip performance to those individuals who do not have chip handling capability. This carrier is also available with a bypassed source resistor preset for maximum linear-power output.

With the FET constructed and packaged, we now can look at its operation. Normally, a positive voltage is placed on the drain relative to the source. This is usually between 2 and 10 V, but can be higher for power FETs. If the gate is first connected to the source ($V_{GS} = 0$) and the drain voltage gradually increased, the source-drain current will increase and then remain almost unchanging. Figure 5.26(a) shows a plot such as would be obtained. If the voltage is increased beyond a certain point (BV_{DS}), a catastrophic failure will occur. This voltage is somewhat higher than the maximum given on the data sheet. The current level at which the current saturates is called I_{DSS}, current-drain-to-source, with gate shorted to the source. In microwave FETs the current saturates because the electrons reach a velocity which is limited by collisions with atoms in the GaAs crystal lattice. Increasing the voltage further simply causes more violent collisions but no increase in current. This velocity is about 1×10^7 cm/s. In lower frequency FETs current saturation occurs in a different way. The gate is too long to allow building up the necessary field to cause velocity saturation, but the voltage drop under the gate will eventually become as large as the pinchoff voltage and saturation will occur.

If we now refer to Figure 5.27 we will see a configuration which should be familiar. It is a combination of Figures 5.23 and 5.24(a). In this situation, electrons travel from the source to drain through the undepleted part of the channel. It should be noted that, even if the external gate-to-source voltage is zero, the gate-channel diodes have a "built-in" voltage equal to about 0.8 V. This means that the channel will always be partially depleted by the built-in voltage, even if no external voltage is applied. Next the gate is disconnected from the source, and a negative voltage is applied to the gate. Figure 5.26(b) shows the characteristic of a FET with 0, –1, –2, –3, and –4 V applied to the gate. As the gate is made more negative, there is less of the channel left to carry current, and the saturated current is lower. Also shown in the

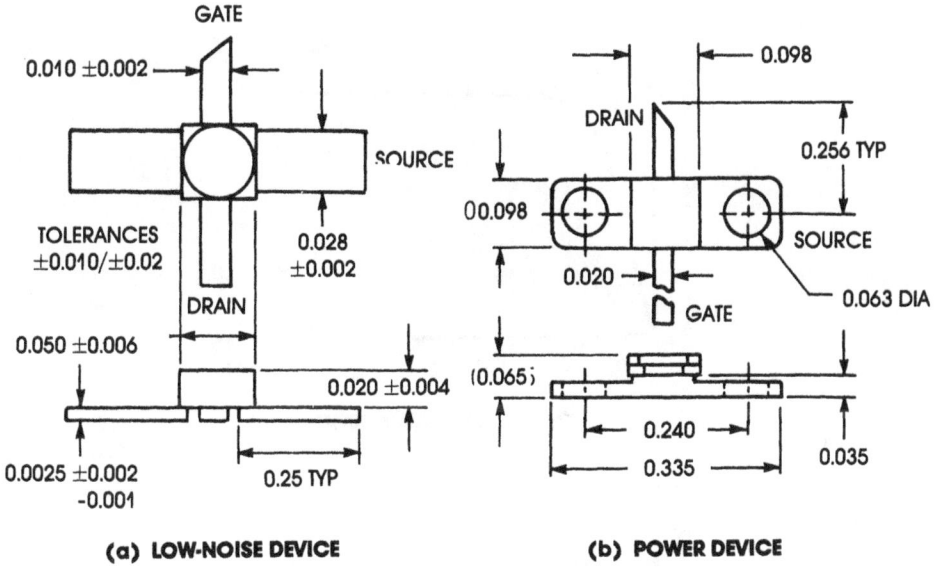

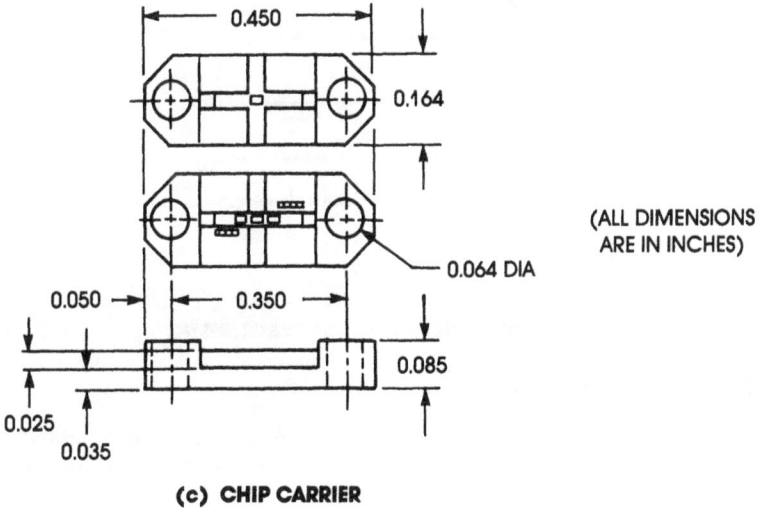

Figure 5.25 GaAs FET Packages

example is the pinchoff voltage, V_p, which is approximately -4 V. Since the FET current never really reaches zero at pinchoff, an arbitrary drain current is selected for defining V_p. For example, define V_p as the gate voltage necessary to reduce the drain current to 1% of I_{DSS}. If $I_{DSS} = 100$ mA, then

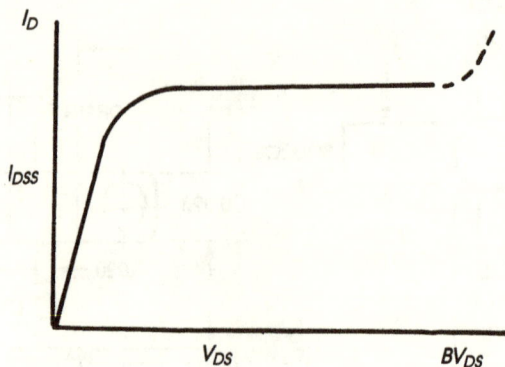

(a) I_{DSS} **CHARACTERISTICS**

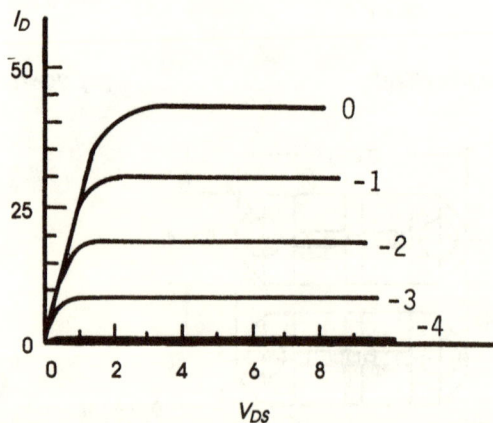

(b) COMPLETE FET CHARACTERISTIC

Figure 5.26 FET Characteristics

V_p is the gate voltage to reduce I_d to 1 mA. As a matter of practicality, V_p is defined by a drain current of 0.1 to 1 mA for small FETs and at larger current for power FETs.

The preceding description of FET action (*dc*) is also applicable to *ac* or RF conditions. First an appropriate positive voltage is applied to the drain. Next a negative voltage is applied to the gate which will give the required drain current. After this, a signal can be applied to the gate. The signal voltage will alternately widen or narrow the depletion layer under the gate, causing the drain to source current also to vary. Note that when the gate is positive, the current is increased, and when negative the drain current decreases. When the drain current is positive and high, the voltage across the load will

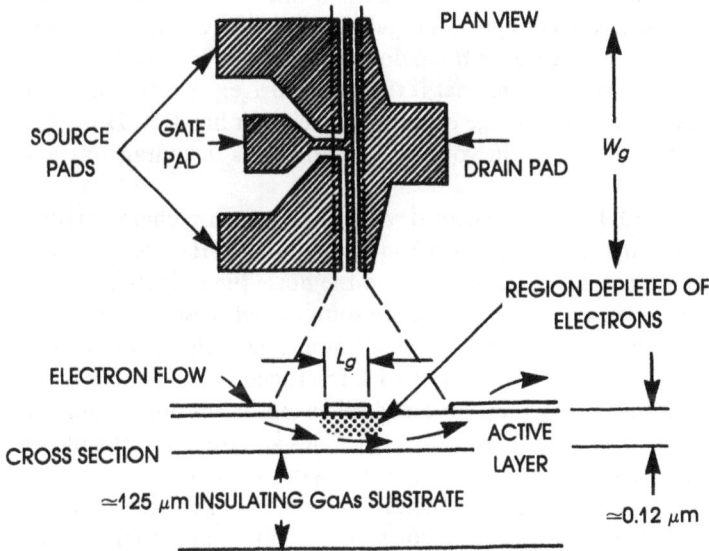

Figure 5.27 Typical Low-Noise GaAs FET

be maximum and the voltage on the drain minimum. Thus, at low frequencies a common-source FET exhibits a 180° phase shift from gate to drain. At higher frequencies this changes, due to reactive elements and internal phase shift.

The gain of a FET is described by a parameter called transconductance. It is defined as

$$g_m = \frac{d\, I_{DSS}}{d\, V_{gs}}$$

For the FET shown in Figure 5.26(b) and g_m is

$$g_m = \frac{0.015}{1.0} = 15 \text{ mmhos}$$

at $V_{gs} = -1$ V. Note that g_m drops as the gate becomes more negative. This is because the channel is getting thinner and its conductance lower. Transconductance is frequency sensitive and drops as frequency is increased.

There are two additional parameters which are also important to the circuit designer. These are the breakdown voltages, gate-to-source and gate-to-drain (BV_{GSo} BV_{GDo}) with the unused terminal unconnected. When measured, these voltages will be almost the same. In an actual circuit, however, it will be noted that the gate-to-drain voltage will be the sum of the gate and drain voltages. The gate breakdown voltages are "soft," that is, the current change near breakdown is gradual. This is very desirable in some circuits, for it means that if the gate is driven slightly into breakdown by a large signal, no damage occurs. On the other hand, if BV_{DS} is exceeded by using too high a drain voltage, breakdown is instantaneous and catastrophic.

GaAs FETs are excellent devices for low-noise characteristics. Their frequency range can be from 0.5 GHz up to 30 GHz. Below 0.5 GHz the $1/f$ noise dominates performance and the noise figure increases.

Since the FET has several noise sources and these are correlated, there is an optimum input tuning for best noise figure which is not the same as that for gain. As a result, when the FET is tuned for best noise figure, some gain will be lost. The gain obtained when optimum noise figure tuning is used is called the "associated gain," G_A. (This and other terms will be covered in the next section, when we present a low-noise data sheet.)

To utilize GaAs FETs to their full potential and choose the right devices for your particular application, you should be familiar with the terms present in a low-noise device data sheet. We will now present these.

The first thing to look for when considering a low-noise GaAs FET is the frequency of operation. If possible, try to choose a device that operates at a higher frequency than needed, so that the portion of the curve that is increasing in noise figure is avoided. For example in Figure 5.28, operating at 8 GHz or less would be needed, if the device were used with the characteristics shown. Above that frequency the noise figure increases, and the associated gain decreases, which is not an ideal situation. So be sure your frequency of operation is below the maximum frequency of the device used. One note on Figure 5.28 is that this curve holds for the condition given — V_{DS} = 3 V, and I_{DS} = 10 mA; that is, a drain-to-source voltage of 3 V and a drain current of 10 mA. If your FET is biased differently than these conditions, it will not perform as shown in Figure 5.28.

The noise figure on a data sheet can be termed optimum noise figure (NF_{OPT}), minimum noise figure (NF_{MIN}), or spot noise figure. Any one of these terms may be used, but they all say that this figure is the lowest noise figure when the conditions listed (V_{DS} and I_{DS}) are used and the corresponding S-parameters for these conditions are used.

Gain is a very important parameter for a low-noise circuit. There are actually two gain numbers involved when talking of low-noise GaAs FETs. One is the maximum available gain (G_{MAX} or *MAG*) and the other is the gain at optimum noise figure (G_{NF} or *Ga*).

The maximum available gain figure is not really concerned with low-noise performance. It is concerned with obtaining the most output available

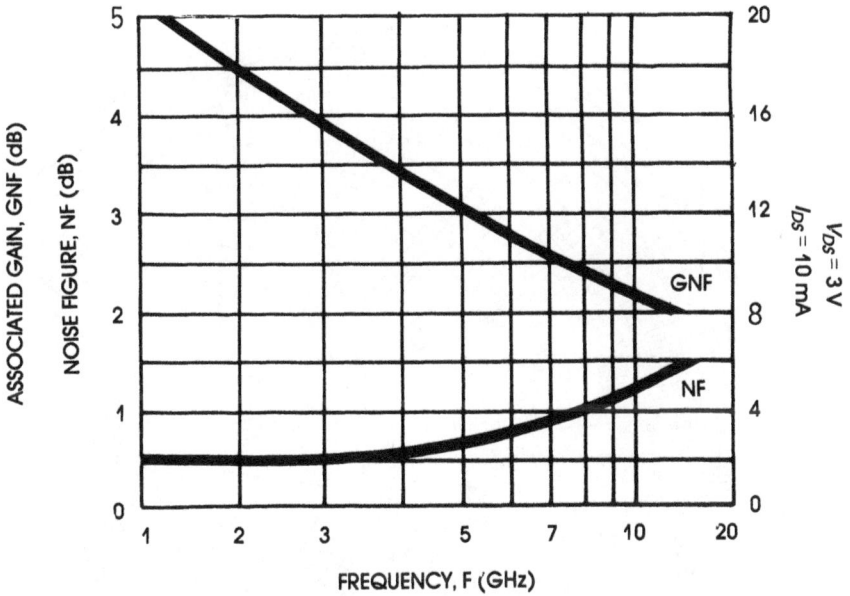

Figure 5.28 Noise Figure and Gain *versus* Frequency

from a device. Conditions are once again important when obtaining this gain figure. The conditions for the maximum gain are usually spelled out as V_{DS} = 3 V, I_{DS} = 30 mA. Notice the greater amount of current necessary to obtain the increased gain.

The gain at optimum noise figure numbers are those obtained when the conditions are duplicated for the best noise figure obtained. That is, we have said that the best noise figure was obtained when V_{DS} = 3 V, and I_{DS} = 10 mA. We, therefore, will duplicate the specified gain when these conditions are met. This gain, as might be imagined, is lower than the maximum available gain, since we are operating with less drain current in order to get a low noise figure. Less drain current will automatically reduce the gain. The gain figure can be on the order of 2 to 3 dB difference from the maximum available gain figure. This difference, however, can be made up in following stages if increased gain is really important. The prime function of the first stage is low noise.

An important part of obtaining low-noise performance is the input and output match of the device: S-parameters. Figure 5.29 shows S_{11} and S_{22} (input and output reflection characteristics, respectively) for a chip device. Figure 5.30 is the S_{11} and S_{22} values for a low-noise packaged GaAs FET device. Test conditions are slightly different, but notice how the packaged

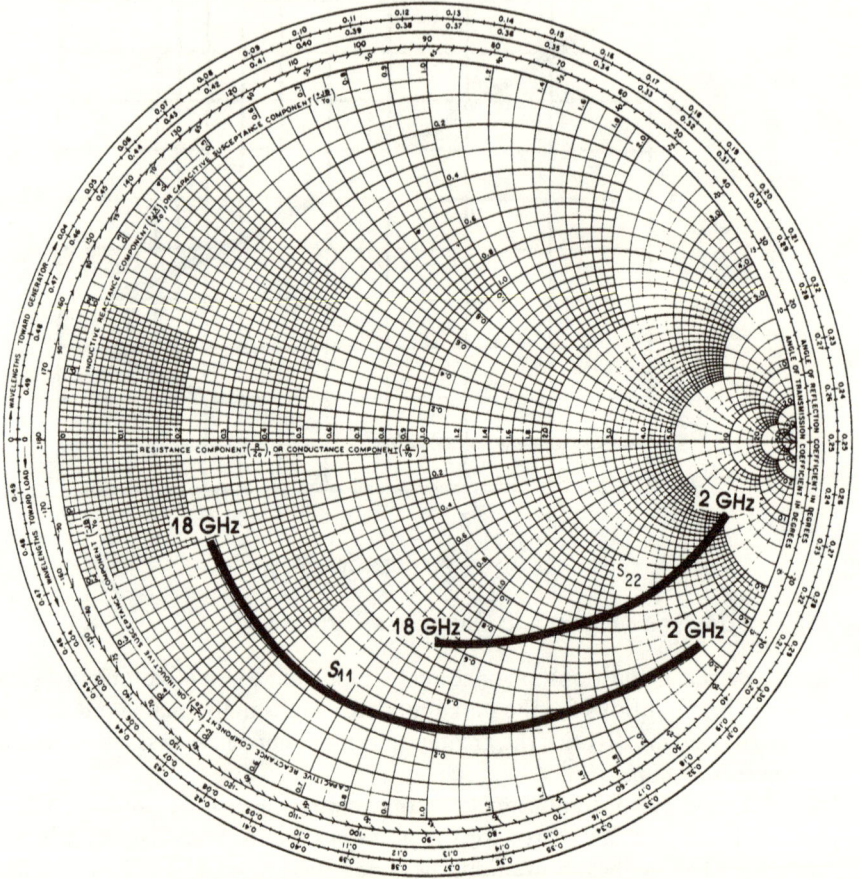

Figure 5.29 S_{11} and S_{22} for a Chip Device

device spreads out S_{11} and S_{22} considerably over the chip version. (S_{11} and S_{22} are the input and output reflection coefficients, respectively.)

Additional parameters which will appear on low-noise device data sheets are shown below. Most have been defined previously.

I_{DSS} — Saturated drain current (maximum)

I_{GS} — Gate to source leakage current

I_{GD} — Gate to drain leakage current

I_{DS} — Drain current

V_{DS} — Drain to source voltage

V_{GS} — Gate to source voltage

g_m — Transconductance (ratio of change in drain current divided by the change in gate voltage as previously defined).

When FETs are made very large, a design such as that shown in Figure 5.24(b) becomes impractical because of the extreme width of the gate. To get around this problem, the gate is divided into segments which are connected in parallel. Figure 5.24 presented a power FET with eight gate segments. When the gate is broken up into segments, a topological problem arises. One of the elements (the drain in this case) becomes enclosed by the other elements and must be brought out by bridging. Metal straps over air (air bridges) or metal straps over silicon dioxide (dielectric bridges) are sometimes used. Figure 5.24 shows wire bridges.

As the signal travels down the gate from the end feed point, the input signal undergoes a loss and phase shift which are worse at higher frequencies. The gate metal resistance combined with the gate-to-source capacitance form what is the equivalent of a lossy RC transmission line. As a result the gate near the feed point and at the far end receive signals which are different in amplitude and phase. To compensate for this, the gate is broken up into narrower and narrower (W_g) segments for higher frequency FETs.

A figure of merit for power FETs is watts per millimeter of gate width. For 4-GHz lab FETs the number is about 1 W/mm; at Ku band this drops to 0.3 to 0.4 W/mm. A conservative operating level for X band is 0.4 to 0.6 W/mm.

As FETs are made increasingly larger to obtain higher power, the matching problem becomes more acute because the input impedance becomes much less than 50 Ω. At higher frequencies the problem becomes even more severe, because the impedance level also drops with increasing frequency. Some of the larger packaged power FETs employ matching or partial matching at the input and output, using inductors and capacitors inside the package. This alleviates the problem considerably but limits bandwidth to a specific frequency range.

Thermal resisitance is important in power FETs as it is in power bipolar transistors. While FETs do not "run away" as do bipolars, the thermal resistance of GaAs is higher than that of silicon. The safe operating region for a FET is bounded by BV_{DS} and I_{DS} at low dissipation and about 180°C channel temperature at high current and voltage, as shown in Figure 5.31. The correct operating temperature should be based on the MTTF desired.

As in the case for low-noise devices, items from a data sheet are presented to acquaint the reader with important terms.

The first parameter to consider, of course, is the frequency of operation. If the short paragraph at the top of the sheet does not call out a specific frequency range, it will usually call out a band. If you are not familiar with the frequency bands, look further down the data sheet and notice frequencies for specific parameters and determine the operating range of the device.

The second parameter to be concerned with is the output power, which will be presented in two ways: power output and power output at the 1-dB

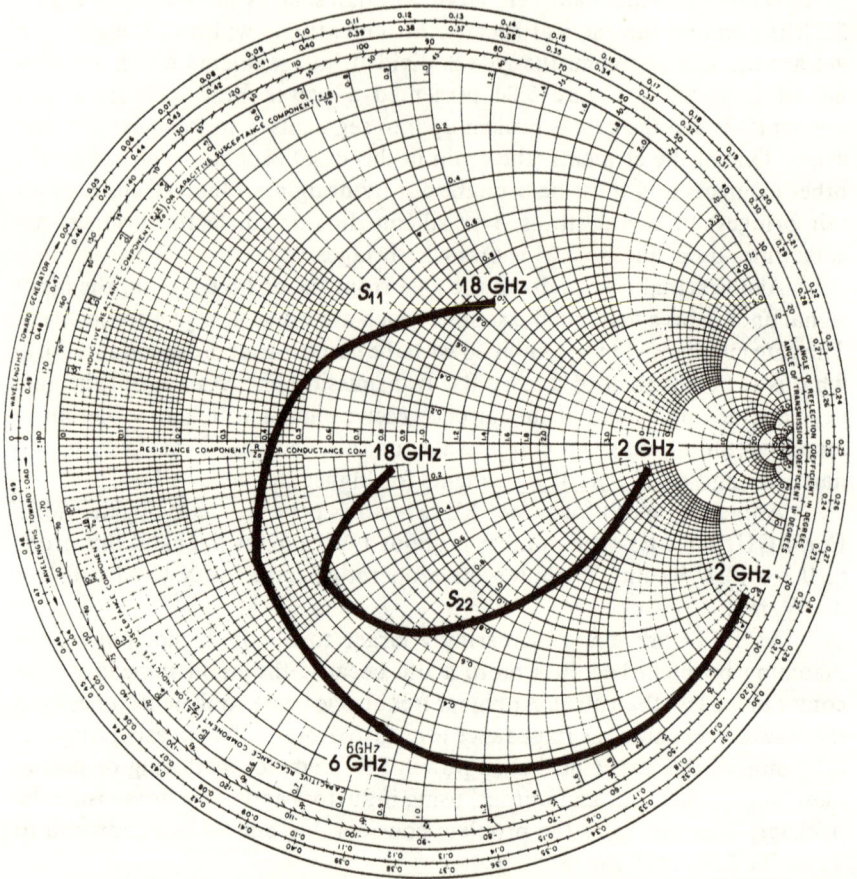

Figure 5.30 S_{11} and S_{22} for a Packaged Device

compression point. The power output figure is the one that tells the linear power output; that is, the power output of the device when operating within a specified input power range. This range does not drive the device into compression. For this power output figure, there must be conditions set down that will result in this power output. These conditions are power input, V_{DS}, and either I_{DS} or V_{GS} (V_{DS} is drain to source voltage, I_{DS} is drain to source current, and V_{GS} is gate to source voltage).

Power output at the 1-dB compression point is found by increasing the power input in 10-dB steps and noting that the output also increases by 10 dB. When a 10-dB input results in a 9-dB increase in output, the 1-dB compression point has been reached. To increase the output further would

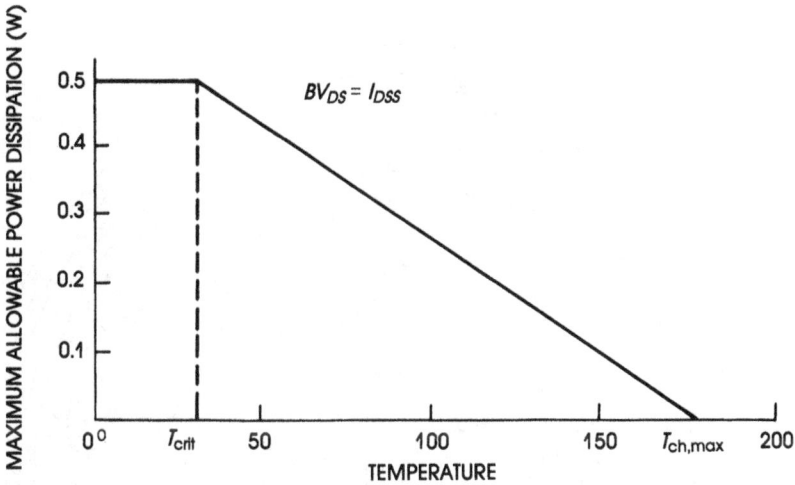

Figure 5.31 Power Dissipation

not make any sense because you will get no more power at the output; the limit has been reached. Many times the 1-dB compression output power is used to keep the output of an amplifier constant over a wide range of environmental conditions or over a varying range of input powers.

Gain, of course, is also an important parameter to consider in a power device. It would be foolish to supply dc and RF power to a device and only produce 1 or 2 dB of gain. Most power FETs will exhibit from 7 to 10 dB of gain up to around 8 GHz with very little effort. Be sure to check the particular device to be used to be sure to know what gain it should produce.

The use of power devices brings to light a whole new set of parameters. These parameters are thermal. First, check the operating temperature range of the device to be used. Most devices will operate from around –65°C to +175°C. These temperatures are somewhat less than the temperatures of their bipolar brothers who will operate up to +200°C. When making the transition from bipolar to GaAs FETs, you should be aware of this temperature difference.

The other thermal parameter, thermal resistance, is expressed in degrees Celsius per watt and tells how well the device gets the heat generated within the device out of that same device. This figure should be as small as possible, since as little rise in temperature per watt of power as possible is wanted.

The final parameter to be investigated is the impedance of the device to be used (S_{11} and S_{22}). Figure 5.32 shows a typical plot of S_{11} and S_{22} for a power device that produces a 1-W output at 8 GHz. Notice how much

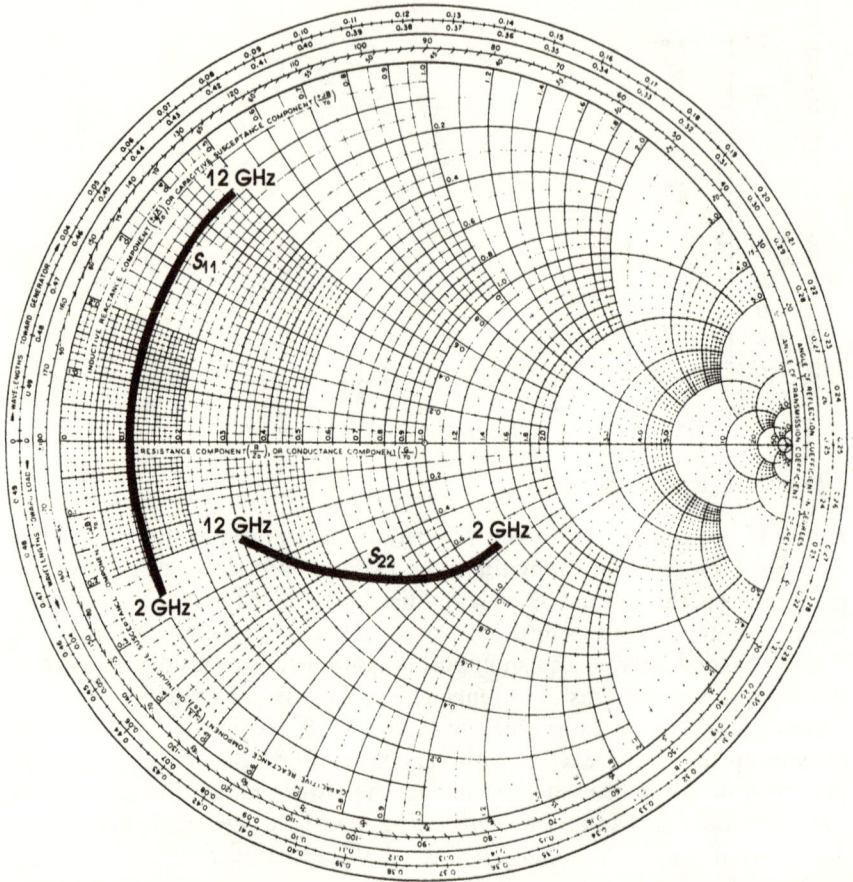

Figure 5.32 S_{11} and S_{22} for a Power Device

lower the impedance is at the input than for the low-noise device. This lower impedance will make the input matching to a 50-Ω circuit a little more difficult and may require multisection matching, if a broad band of frequencies is to be covered.

One area that is of prime importance, but is not covered on data sheets, is handling precautions for GaAs FETs. These precautions are for both low-noise devices and power FETs.

When handling the packaged unit, never pass the device from one person to the next by the leads, especially on a dry day, since static charges will destroy it. Although the MESFET is not as sensitive as a MOSFET, it is still quite delicate.

When soldering the device into a circuit, the soldering iron must be properly grounded. Soldering time should not exceed 20 s at 260°C. Never insert a FET into a prebiased circuit, particularly one where self-biasing is used; the bypass capacitor will charge through the forward-biased barrier and most likely will destroy the gate.

Once the device is soldered into the circuit, do not just "switch on" the gate and drain bias supplies. Many regulated power supplies feature large in-rush transients due to their stability-limited gain-bandwidth product. Adjust V_{GS} slowly to about −1.0 V, gradually increase V_{DS} from 0 V to the desired bias point and, finally, readjust V_{GS} to give the desired drain current.

Remember, for a FET, $V_{DG} = V_{DS} - V_{GS}$, so place no more than 10 V between the drain and gate. If something is suspect, avoid the use of a digital multimeter to check the resistance because the voltage supply within the meter may destroy the gate.

We must also caution against the use of curve tracers. The high-voltage transformers used in most curve tracers may have high leakage currents that can, if not properly grounded, destroy the FET. If a curve tracer test is required, we suggest the following:

1. Ground the curve tracer to earth.
2. Set the gate-source voltage to zero voltage.
3. Increase the drain-source voltage slowly to the desired value ($\approx$3 V).
4. Adjust the gate-source voltage to the desired value.

When handling GaAs FET chips, the most important precaution to remember is that all operations must be performed in a clean environment, by the fewest number of people possible. If incoming inspection is performed, leave the chips in the carrier and store in a dry environment, preferably in dry nitrogen. When removing the chip from the carrier, use a vacuum probe with a polytetrafluoroethylyne (PTFE) tip. If tweezers are used, extreme caution is mandatory. Gallium arsenide is more brittle than silicon, and dust may break off and become lodged in the small channel.

The back of the chip is metallized with pure gold and can be die-attached with low-temperature epoxy, AuGe or AuSn 0.5-mil preforms. Whatever material is used, the maximum allowable exposure is 300°C for five minutes. The use of a die-attach machine is not recommended because the gate may be damaged.

If AuSn preforms are used, recommendations are heating the stage to 200°C and the probe to 250°C. Probe pressure should be less than 22 g. For specific chip handling procedures, consult the individual manufacturer for any peculiarities that may be present for their devices.

In general, three simple precautions will prevent the destruction of devices from transient breakdown:

1. Unless it is to be negatively biased, dc-ground the device gate.
2. Decouple the bias ports near the device with a low-inductance capacitor (1 μF tantalum) shunted with a zener diode (5.8 V, 1.3 W). In addition to limiting transients, the zener diode gives protection against over-voltage and reverse biasing.
3. Switch on power supplies and set their outputs to minimum before connecting the device.

Thus, two types of GaAs FETs have been presented. It can be seen how they are matured devices for microwave operations. Their structure and operations make them more and more suitable for microwave and millimeter application.

5.3 HIGH ELECTRON MOBILITY TRANSISTORS (HEMT)

The latest addition to the lineup of solid-state devices designed primarily for microwave applications is the high electron mobility transistor (HEMT). This is a relatively new technology, at the time of this text, and is going through the typical growing pains that both bipolar and GaAs devices have been through. When they were first proposed in 1978, they received a rather indifferent type of reception. They have, however, recently attained a much higher credibility in the microwave industry by demonstrating their performance with commercially available devices. Tests have shown some areas where HEMTs have outperformed GaAs FETs. It should be pointed out here that the high electron mobility transistor is designed to be a low-noise device only. It is not designed to compete with high-power GaAs FETs at microwave frequencies.

The HEMT device looks very similar to the GaAs FET. Figure 5.33 shows a typical geometry for a HEMT. Notice its striking similarity to the low-noise GaAs FET discussed in the previous section. Also, the structure is shown to be similar as presented in Figure 5.34. These two figures lead one to believe that nothing is really different about the HEMT, so why does it perform as well? The answer is not in the structure, as we have seen, but rather the semiconductor materials it uses. These devices are based on the modulation-doped GaAs/AlGaAs hetrostructure epitaxy in which the motion of the charge carriers is confined to a thin sheet within the GaAs buffer layer. By referring once again to Figure 5.34 you can see the layers being discussed.

In the HEMT, a silicon-doped AlGaAs layer grown on top of an undoped GaAs layer brings about the formation of a two-dimensional electron gas on the GaAs side of the heterostructure (Figure 5.34). The gas forms because of the electrons' greater affinity to the GaAs. The two-dimensional gas layer is about 150Å thick and forms the carrier channel that links the device source and drain.

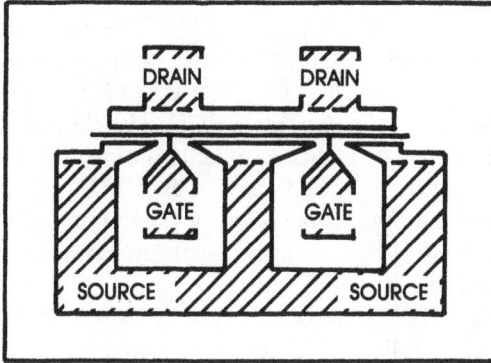

Figure 5.33 HEMT Geometry

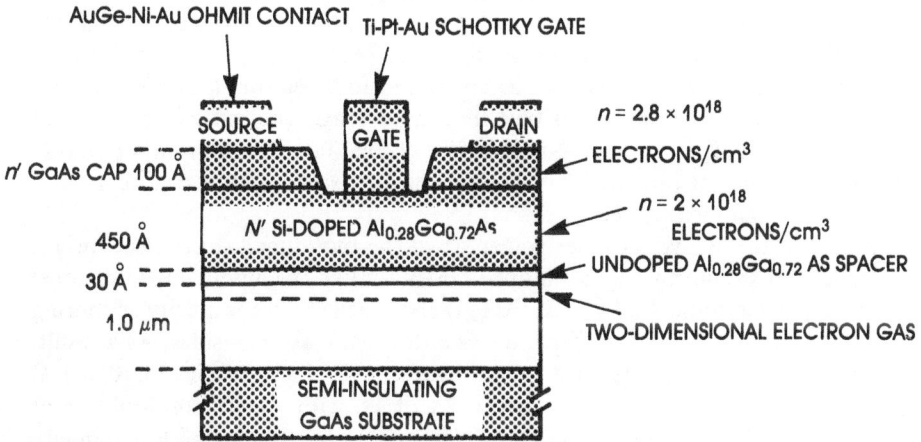

Figure 5.34 HEMT Construction

When a Schottky barrier gate is placed on top of the AlGaAs layer, a depletion region forms beneath the gate. If the AlGaAs layer is sufficiently thick, the gate and the interface depletion regions will not overlap, and the device will operate as a "normally on" transistor. In this *depletion-mode* device, an application of negative bias to the gate will extend the gate depletion region to the hetero-junction interface, thereby barring electron flow and pinching off the drain-source current.

Etching a recess to seat the gate deeper in the AlGaAs layer creates a device that operates in the *enhancement mode*. Such a device operates as a "normally off" transistor. When a positive voltage greater than the threshold voltage is applied to this type of HEMT, electrons will accumulate at the hetero-junction interface and form the two-dimensional gas, turning the device on. Thus, unlike the MESFET, the HEMT actually functions more like a metal oxide semiconductor (MOSFET), with the Schottky barrier gate controlling the number of electrons in the two-dimensional electron gas by raising and lowering the interface barrier. In a MESFET, bias applied to the gate modulates the channel and thereby the number of electrons passing through it. With a HEMT, the channel thickness remains constant; it is the number of carriers that is modulated.

Electrons in a HEMT move in a very thin sheet of material, traveling in a direction that is parallel to the hetero-junction interface, with coordinates x and y (two-dimensional motion). In a MESFET, electrons are allowed three degrees of movement: x, y, and z (three-dimensional motion).

There is no actual proven connection between the degree of electron freedom and a device's noise figure, but certain theories have arisen that the HEMT has a lower noise figure than the MESFET because of the two-dimensional electron freedom as compared to three for the FET. To substantiate these theories, it is true that the average quadratic fluctuation of the electron drift velocity — which is a critical factor in determining noise as a measure of random electron motion — is smaller with two degrees of freedom than with three.

In the HEMT, carrier transport in the two-dimensional electron gas is similar to movement in undoped GaAs. Unlike the highly doped channel of a conventional GaAs MESFET, there is little or no impurity scattering in the undoped GaAs in which the two-dimensional gas resides. As a result, electrons in HEMTs travel at twice the saturated velocity (about 2×10^7 cm/s) at room temperature as those in GaAs FETs, with an electron mobility of nearly 8000 cm^2/V-s (compared to 4000 cm^2/V-s for the GaAs FET channel). At lower temperatures, 77 K for example, the electron velocity increases to about 3×10^7 cm/s, while the mobility increases to approximately 80,000 cm^2/V-s, depending on the epitaxial layer structure.

HEMT fabrication involves a multistage process that includes molecular-beam epitaxy and ion-implantation techniques. Once the doped and undoped layers are created, a wafer undergoes a succession of processes that produce electrical isolation, establish ohmic contacts, and form gates and gate recesses.

Both the isolation and ohmic-contact wafer processing steps utilize photolithographic methods, which are effective with features of 1 μm or larger. In order to form smaller gates, such as 0.25 μm, electron-beam lithography is employed. In this process, a computer-controlled beam of electrons inscribes the features directly onto the wafer, without a mask. E-beam lithography is capable of producing minimum features of 0.15 μm, though still with limited yields.

We have been making general comparisons of GaAs FETs and HEMTs throughout this section. Now let us show some data. Table 5.3 is an actual test run to compare the two devices side-by-side.

Table 5.3

Parameters and Conditions	*GaAs FET*	*HEMT*
Minimum noise figure (dB) at V_{DS} + 3.5 Vdc and I_{DS} + 10 to 15 mA		
at 4 GHz	0.7	0.4
at 8 GHz	1.4	0.8
at 12 GHz	1.9	1.2
at 18 GHz	2.6	1.8
Associated gain (dB) at V_{DS} + 3.5 Vdc and I_{DS} + 10 to 15 mA		
at 8 GHz	11.5	12.0
at 12 GHz	9.5	10.5
at 18 GHz	7.5	9.5
Maximum available gain (dB) at V_{DS} + 4 Vdc and I_{DS} + 0.5 I_{DSS}		
at 4 GHz	17.0	17.0
at 8 GHz	13.0	15.0
at 12 GHz	11.5	14.0
at 18 GHz	9.5	11.5
Output power at 1 dB compression (dBm) at V_{DS} + 4 Vdc and I_{DS} + 0.5 I_{DSS}		
at 12 GHz	13.0	13.0
Saturated drain current (mA) at V_{DS} + 3 Vdc and V_{GS} + 0 Vdc	45	35
Transconductance (mmho) at V_{DS} + 3 Vdc and V_{GS} + 0.2 Vdc	40	45

Note: All specifications are at +25° C.

It can be seen that for these particular devices the HEMT has either matched or surpassed the FET in all categories. A second set of tests has yielded Figure 5.35. This shows maximum available gain and noise figure

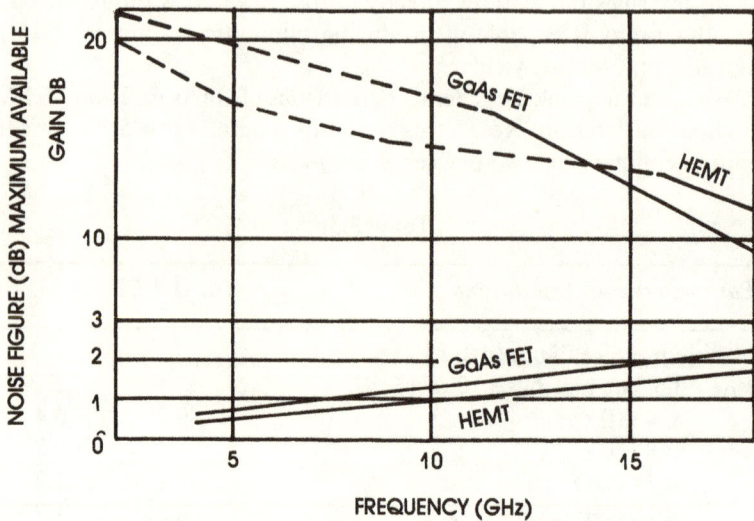

Figure 5.35 GaAs FET *versus* HEMT

for a 0.5-μm gate HEMT and a 0.3-μm gate GaAs FET. It can be seen that the GaAs FET outdoes the HEMT in available gain up to about 14 GHz, but the HEMT is superior in noise figure throughout.

With its inherent electron speed advantages, the HEMT has naturally generated great optimism among digital circuit designers. HEMT ring oscillators have already been fabricated with switching delays of only 12 ps at room temperature.

Only the Josephson junction exhibits lower power dissipation per gate than the HEMT, 4 K (compared to 77-K operation for the HEMT). Research on the application of the HEMT as a digital device continues.

The HEMT also brings optimism to analog designers, especially those working at millimeter-wave frequencies. With its low noise and high gain, 0.2-μm HEMTs may one day form the heart of low-noise amplifiers working up to 100 GHz. Such transistors have been fabricated at TRW and are currently undergoing full evaluation. In the meantime, HEMTs with gate lengths ranging from 0.50 to 0.25 μm have completed testing by a number of manufacturers and show good performance up to 45 GHz. (An "average" HEMT, is one having a 0.35-μm gate length. Such a transistor has a nominal gate width of 75 μm and dc transconductance of 240 mS/mm of gate periphery.)

While the HEMT should see the majority of its analog use at millimeter-wave frequencies, the technology has also been explored for lower-frequency applications. All of the characteristics that make the HEMT such a viable candidate for millimeter-wave work also make it attractive for low power-consumption, lower frequency designs.

Thus, the HEMT (High Electron Mobility Transistor) joins its predecessors in the microwave field. It has had its triumphs to this point and it has had its defects. These are all ups and downs that the bipolar and the GaAs FET have been through and, to some degree, are still going through. The HEMT has found many application areas and will continue to find even more.

5.4 SUMMARY

It was stated at the beginning of this chapter that the microwave transistor is the device which has probably advanced the most of any microwave devices over the years. After proceeding through the chapter I am sure you will agree with that statement. The solid-state transistor as used in microwaves has come to the bipolar, field-effect, and HEMT, each in its own time to meet the needs of that time. Yet, all of the devices are still finding their own application areas today. The microwave transistors — individual devices which work together for better microwave systems.

Chapter 6
Solid-State Components

In all of the previous chapters of this text we have presented either the theory behind the operation of microwave solid-state devices or the devices themselves. This chapter will tie all of this information together by presenting applications of the previously covered theory. Obviously, the theory would be of little use if it was only theory. To make it useful we must produce components which will perform certain specified functions.

As we stated previously, microwave solid-state devices will perform a variety of tasks. In particular, we stated how a diode in the microwave range is capable of amplifying, oscillating, mixing, detecting, attenuating, or switching. All of these functions from a tiny two-element device. This indicates that many functions previously reserved for transistors and tubes are now delegated to the microwave diode.

Similarly, the microwave transistor used to be something that ceased operation around 5 Ghz. But, as we have seen in previous chapters, transistors are available which will operate well past 30 GHz with great reliability and repeatability. Thus, when the microwave solid-state devices are used in microwave components you have capabilities which are almost endless.

6.1 MICROWAVE AMPLIFIERS

When most people think of microwave, or any frequency range, solid-state devices, the first thing that comes to mind is the transistor and its use as an amplifier. Although amplifiers are available with a variety of power outputs, noise figures, and linearities, the ones usually thought of are the low-level, low-noise amplifiers. Generally, these are assumed to be the most understandable and most commonly used types of amplifiers. This is not always the case, but we will assume it is and proceed with our discussions on this assumption.

When we speak of low-level microwave amplifiers we are generally speaking of the amplifier which is directly connected to a receiving antenna or to a filter which is directly connected to an antenna. The position of the amplifier dictates that it must exhibit a low noise figure in order for the entire receiver to operate at its prescribed levels without interference from internal noise. With such a criterion in mind we will proceed to do a general low-noise amplifier design.

The design procedure to be used is one which is typically used for single-stage amplifiers. It consists of matching both the input and output to a 50Ω by means of a transformer-open stub combination. This is a very quick and efficient means of matching a transistor and obtaining the desired properties. Design goals for this amplifier are:

- Frequency   12 GHz
- Noise Figure   <3.5 dB
- Gain   7.5 dB
- Output VSWR   <1.5:1

Note: The gain figure noted is the gain when the amplifier has a noise figure of 3.5 dB or less.

For the amplifier to operate as a low-noise device we must concentrate on its noise parameters rather than on the gain (or power) parameters. Thus, the first criterion in choosing a transistor (after you determine that it will operate in the proper frequency range) is what its rated noise figure is and if the input reflection coefficient (S_{11}) is given to you for minimum noise. The value of S_{11} for a maximum gain is different from that for minimum noise. This is because the operating parameters (V_{CE}, I_C for bipolars; V_{DS}, I_{DS} for FETs) are different. The collector (or drain) current is usually substantially less when using a device for minimum noise. This lower current will result in different S-parameters throughout the device.

The device for our application exhibits the following parameters:

- S_{11} = $0.762 \angle 135°$
- S_{21} = $1.305 \angle -55°$
- S_{12} = $0.079 \angle -85°$
- S_{22} = $0.592 \angle 171°$
- Ga_{max} = 9.2 dB
- Γ_{MS} = $0.861 \angle -128°$
- Γ_{ML} = $0.701 \angle -145°$
- F_{min} = 2.7 dB
- Ga = 8.3 dB
- Γ_0 = $0.73 \angle -140°$
- Γ_L = $0.568 \angle -150°$

Terms that may need further explanation are:

1. Ga_{max} = Maximum available gain from the device when it is matched for gain

2. Γ_{MS} = Source match for maximum available power
gain

3. Γ_{ML} = Load match for maximum available power
gain

4. F_{min} = Minimum noise figure

5. Ga = Associated gain when you have a minimum
noise figure

6. Γ_0 = Optimum source reflection coefficient

7. Γ_L = Optimum load reflection coefficient

To begin with we will look at the input matching network. To accomplish this we must transform our 50-Ω generator to Γ_0 (optimum source reflection coefficient), since the input matching network must appear as to achieve the proper matching of the generator and the transistor. *Note*: We will assume that there is a certain bonding ribbon inductance (0.1 nH). This will have to be subtracted from the impedance value used to determine the matching elements. This impedance value is $X_L = 2\pi f L$ where $f = 12$ GHz and $L = 0.1$ nH. $X_L = 7.5$.

By substituting values into the noise figure impedance equation:

$$Z_{NF} = \frac{(1 - |\Gamma_0|^2)}{1 + |\Gamma_0|^2 - 2|\Gamma_0|\cos\angle\Gamma_0} + j\frac{(2|\Gamma_0|\sin\angle\Gamma_0)}{1 + |\Gamma_0|^2 - 2|\Gamma_0|\cos\angle\Gamma_0}$$

We obtain

$$Z_{NF} = 8.75 - j\,17.23\,\Omega$$

When we adjust for the 0.1 nH inductance, the corrected impedance

$$Z_{NFl} = 8.75 - j\,17.23 - j\,7.5$$
$$Z_{NFl} = 8.75 - j\,24.73\,\Omega$$

We now need to determine Y_{NF}, in order to be able to work with the matching circuit. This can be done mathematically or by plotting on the Smith chart. Figure 6.1 is the Smith chart representation. First, Z_{NFl}, is plotted after being normalized (divided by 50Ω) to the 50-Ω system. This value is 0.175 - j0.49, (Point A). This is then transformed back across the chart until the same distance from the center is reached. This is Y_{NFl}, (Point B). The value real is 0.63 + j1.78. When this is multiplied by 1/50 (0.02) the actual value of Y_{NFl} is 0.0127 + j0.0356. This is an important value since the transformer and stub values depend on Y_{NFl}. With Y_{NFl}, available, we are now able to calculate the quarter-wave transformer. This transformer moves the 50-Ω source impedance (center of the chart) on to the real axis (horizontal line at the center of the chart) and around the conductance circle

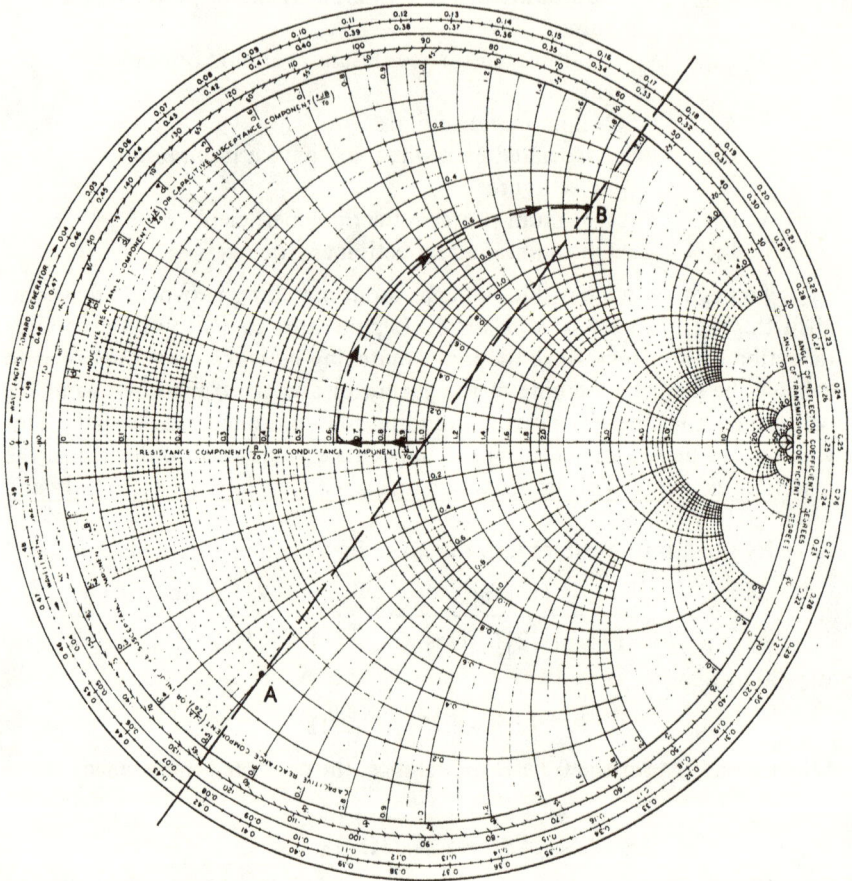

Figure 6.1 Input Match

until we reach Y_{NFI}, (Point B). The value of this transformer is

$$Z_1 = \sqrt{50\left(\frac{1}{\operatorname{Re}Y_{NFI}}\right)} = \sqrt{50\left(\frac{1}{0.0127}\right)}$$
$$Z_1 = 62.75\Omega$$

(The movement described in the paragraph above can be seen in Figure 6.1). This amplifier circuit is to be built of microstrip with a dielectric constant of 10.2. This may be ceramic-loaded PTFE material or aluminum (pure ceramic material). It must be a high dielectric constant so that the widths and lengths of the lines are reasonable. Using ceramic loaded PTFE ($\epsilon_r = 10.2$)

the impedance of 62.75Ω results in a line that is 0.014 inch wide. The length of the line, as dictated by its name quarter-wave transformer, is one quarter-wavelength at 12 GHz. This is calculated as follows:

$$\lambda/4 = \frac{c}{4f\sqrt{\epsilon_{eff}}}$$

where:

c = velocity of light = 3×10^{10} cm/s
f = frequency in Hertz
ϵ_{eff} = the effective dielectric constant for the impedance used (this value is 6.55 for 62.75)

$$\lambda/4 \ = \frac{3 \times 10^{10}}{4(12 \times 10^9)\sqrt{6.55}}$$

$$\lambda/4 \ = 0.244 \text{ cm} = 0.096 \text{ in}$$

Thus, the quarter-wave transformer section is 0.096″ long and 0.014″ wide (as shown in Figure 6.2)

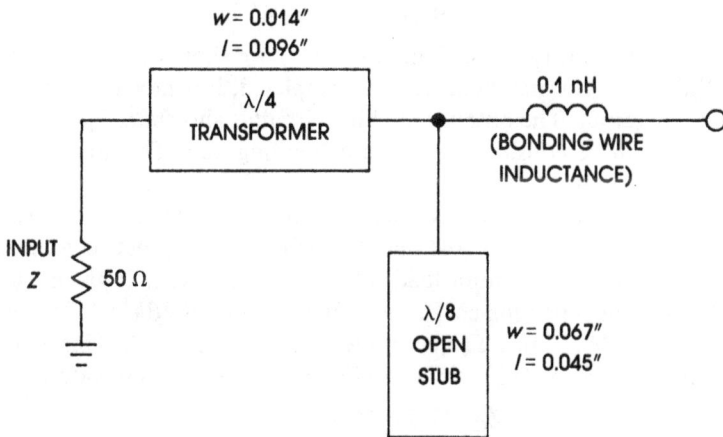

Figure 6.2 Input Matching Network

The second element referred to in the beginning of our discussion and shown in Figure 6.2 is the one-eighth wavelength open stub. To determine how long this one-eighth wavelength is, we must first determine the impedance of the stub. This can be found by going back to Figure 6.1 and reading the imaginary portion of the admittance. This value (which is normalized) is 1.84. This open-circuited stub is capacitive with a value B of 1.84 times 20 mmhos ($1/50\Omega$).

$$B = 1.84 \ (20 \text{ mmhos})$$
$$B = 36.4 \text{ mmhos}$$

The impedance of this stub is $1/B$:

$$Z = \frac{1}{36.4} \text{ mmhos}$$

$$Z = 27.47 \text{ ohms}$$

When we refer back to our values of impedance we find that 27.47Ω is a line which is 0.067 in wide. Also, the effective dielectric constant for this impedance is 7.537. This will be used to calculate the eighth-wavelength as follows:

$$\lambda/8 = \frac{C}{8f\sqrt{\epsilon_{\text{eff}}}}$$

$$\lambda/8 = \frac{3 \times 10^{10}}{8(12 \times 10^9)\sqrt{7.537}}$$

$$\lambda/8 = 0.1138 \text{ cm} = 0.045 \text{ in}$$

Both the dimensions ($w = 0.067$ and $l = 0.045$) are shown in Figure 6.2.

With the input matching circuits completed, it is now time to concentrate on the output. The output matching circuit is shown in Figure 6.3. The first element to be considered will be a bonding pad. This pad is used to give the fabricator room to solder the lead of the transistor, or place a bonding wire if a chip device is used. This length is generally between $10°$ and $20°$ of rotation along a 50Ω line. We will go directly between these and rotate $15°$. The Γ_L (optimum load reflection coefficient) is given as 0.568 $\angle{-150°}$. Since the reflection coefficient angle rotation is $2\beta l (l = 15°)$, the total connection is $30°$. Thus, Γ_0 (connected) is 0.568 $\angle{-120°}$. This makes Z corrected and Y corrected as follows (Figure 6.4 shows these values):

$$Z_C = 17.5 - j\,26 \text{ ohms}$$
$$Y_C = 0.018 + j\,0.0266 \text{ mhos}$$

When determining the length of the pad we will use ceramic-loaded PTFE material. This total correction results in a change of $1/12\lambda$. At 12 GHz this is 0.031 in long. (The width of the 50-Ω is 0.023 in.)

The next element to be considered is the transformer section. This will match the 50-Ω output impedance to the constant conductance circle of Y_C (Figure 6.4). To allow tuning of the amplifier a narrow line is used which

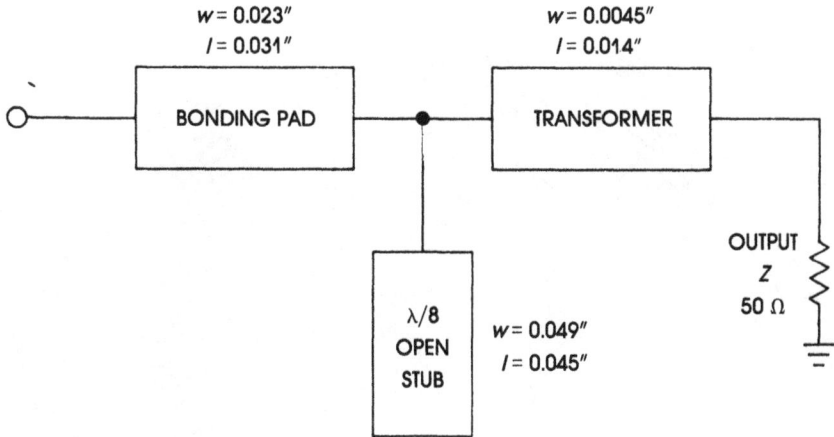

Figure 6.3 Output Matching Network

is meandered so that bonding wires may be attached across it. For our designs we will choose a 90-Ω line (width = 0.0045 in and ϵ_{eff} = 6.245). We must now determine the length of the line (transformer) by using the real part of the admittance calculated previously (G = 18 mmhos): Y_0 = 1/Z_0, which is 1/90 = 11.1 mmhos; and Y_L which is equal to 20 mmhos (1/50 Ω). Using those numbers and the equation below we determine the value of Θ.

$$\text{Tan } \Theta = \frac{Y_0^2 \, (Y_L - G)}{Y_L \, (G Y_L - Y_0^2)}$$

$$\Theta = 12.8°$$

The corresponding length for this Θ is 0.0356 cm or 0.014 in. The admittance which has been transformed is as follows:

$$Y_B = \frac{Y_0 \, (Y_L + j \, Y_0 \, \text{Tan } \Theta)}{Y_0 + j \, Y_L \, \text{Tan } \Theta}$$

$$Y_B = 18.7 - j \, 3.12 \text{ mmhos}$$

(This is shown on the Smith chart in Figure 6.4 as Y_B.)

   The final element is the open stub which will match the reactive portion of the circuit. This is an eighth wavelength stub. Total (B_T) is equal to the corrected value (B_C) minus the transformed value (B_B). That is

$B_T = B_C - B_B$
$B_T = 26.6 - (-3.12)$
$B_T = 29.72$ mmhos

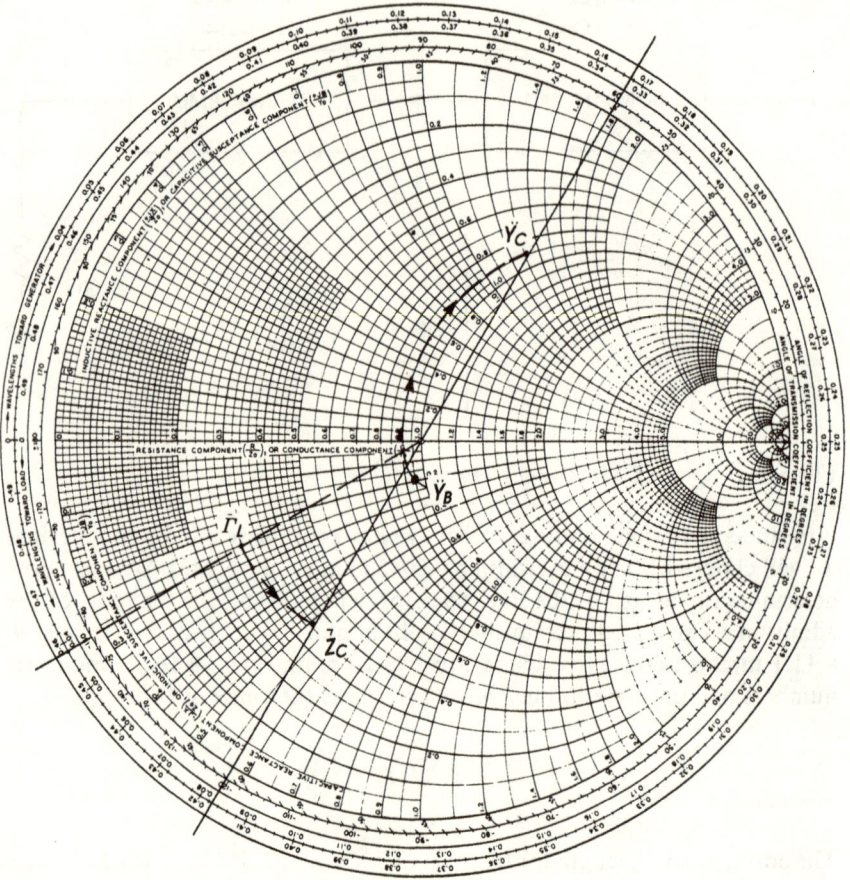

Figure 6.4 Output Matching Circuit

and

$$Z_0 = 1/B_T = 33.6\Omega$$

The width of a 33.6-Ω transmission line in ceramic-loaded PTFE is 0.0490 in. The length of a $\lambda/8$ stub with an impedance of 33.6Ω is calculated to be 0.045 in. The final circuit configuration, without bias circuitry, is shown in Figure 6.5. When the circuit was built it exhibited the parameters shown in Table 6.1 as compared to its initial requirements.

Thus, the parameters required for proper operations have been met.

This amplifier is only one of many that can be designed. Amplifiers can be designed to be used at low power levels, medium power levels, high power levels, low-noise, or linear operation over a specified range of input

TABLE 6.1

Parameter	Requirement	Measured Data
Frequency	12 GHz	12 GHz
Noise Figure	<3.5 dB	3.2 dB
Gain	7.5 dB	7.6 dB
Output VSWR	<1.5:1	1.22:1

level. Whichever type of amplifier is used, you will need to know the porperties and characteristics of the device to be used and operate with them in much the same way as we have for this example.

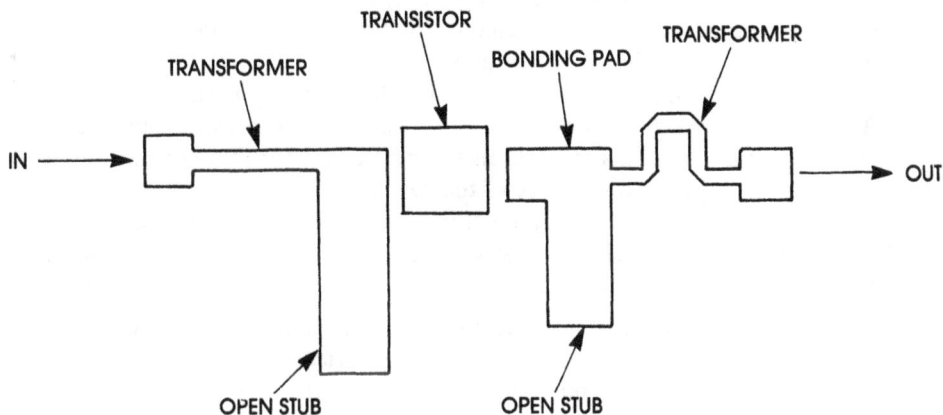

Figure 6.5 Completed Amplifier

6.2 MICROWAVE OSCILLATORS

Probably the second most prominent devices we think of when considering solid-state components are oscillators. We will look at two types of commonly used microwave oscillators: transistor and Gunn oscillators.

Whenever designers set out to produce an oscillator circuit, they always seem to hear the old syaing: "If you want an oscillator, design an amplifier. It will surely oscillate." This saying, although rather disturbing to the new and inexperienced designer, has a lot of merit to it. This is because, generally, an oscillator is built around an amplifier configuration. To illustrate this let us look at one method for designing a bipolar transistor oscillator. This method is one which has small signal S-parameters.

To have a stable oscillation in a circuit, the circuit requires a load impedance whose real and imaginary parts are equal in amplitude and opposite in sign to the impedance of the transistor used. Diagrams are used which will simulate this condition when the transistor S-parameters are put into the computer. The problems with these programs arise when we consider the last term we used — S-parameters. These parameters are dependent on bias and power-level conditions. If these are not exactly as presented in the computer simulation, the circuit will satisfy all conditions for oscillations, but will not oscillate. This is because the negative resistance designed does not match the load and oscillations cannot be sustained.

To eliminate such a problem the designer can design the negative resistance magnitude to be larger than the load resistance. Usually 2 1/2 to 3 times larger will do the job. This will enable the circuit to increase power until the proper resistance is met. This will, however, also give a different reactance component which will change the frequency. The designer should be aware of this and design accordingly.

So it would appear that there must be a better way to design oscillators. A two-step process has been suggested and used which provides stable and reliable oscillations. This involves installing a transistor in a nonoscillating circuit, measuring the impedance of the device over a range of input power, and using these data to design a matching circuit which will cause the circuit to oscillate. Figure 6.6 shows what we are referring to in our explanations. Steps to be taken in such a design are:

a. Place the transistor in the nonoscillating circuit (Figure 6.6).
b. Measure the impedance (Z) of the device over a range of input powers. (Record the reflected power and the power levels.)
c. Subtract the input (incident) power from the reflected power. This is called the *added power*.
d. Make note of the impedance which is measured at the power level corresponding to maximum *added* power.
e. Design the matching circuit (Figure 6.6) which exhibits an impedance equal to the negative of the impedance determined in step d above.

To illustrate this procedure let us do a straightforward design of a bipolar 4.5-GHz oscillator. An *npn* bipolar device is to be used, V_{CE} = 15 V, and I_C = 30 mA. Tests were run using 50-Ω lines with transmission lines off both the emitter (open circuit) and base (shorted lines). The test resulted in a maximum added power of 19 dBm and a normalized impedance value of 0.27 + j0.35 (13.5 + j17.5Ω). This is shown as point A in Figure 6.7. To transform this impedance to a 50-Ω load we first must provide a shunt line that has susceptance equal to point B in Figure 6.7. This is the point on the chart where the VSWR circle intersects the unity circle (where the normalized characteristic impedance, Z_0, is equal to 1.0). This point is an open circuit

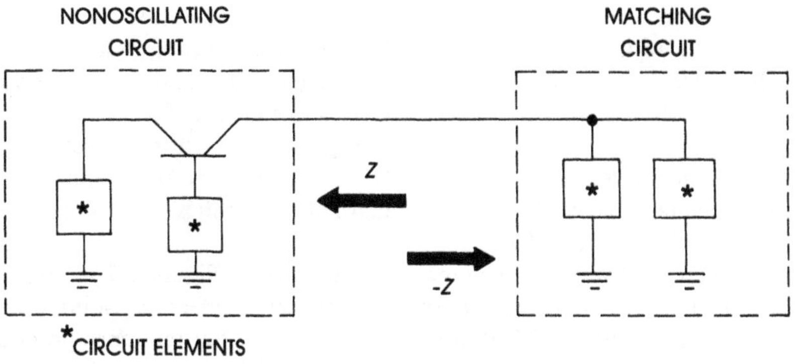

Figure 6.6 Oscillator Circuit

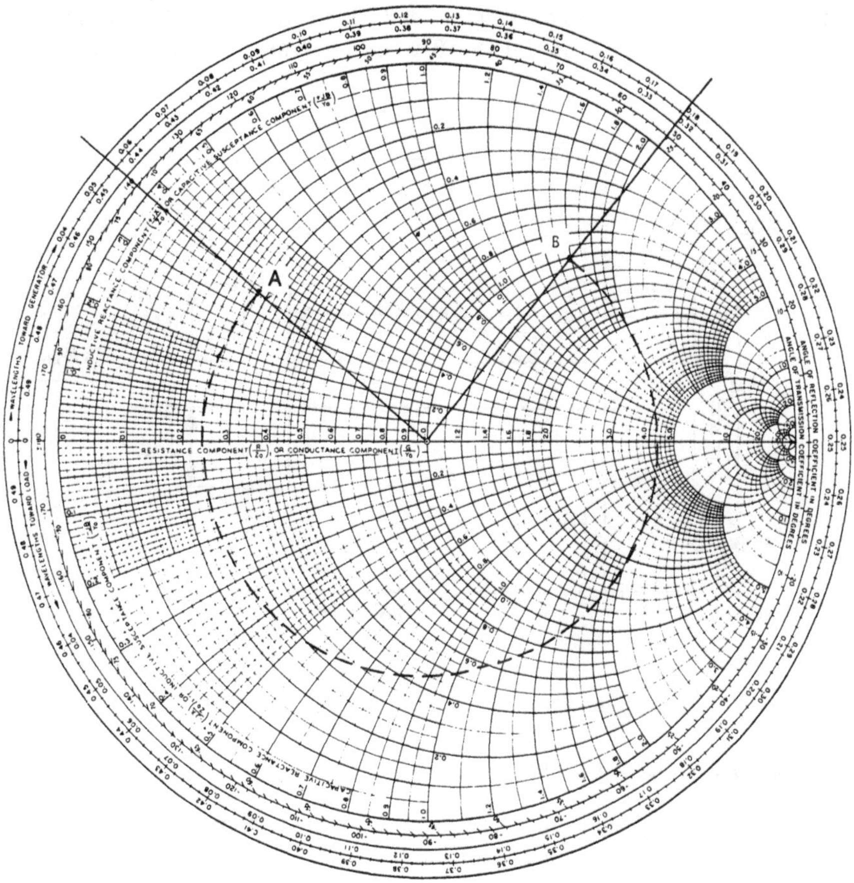

Figure 6.7 Oscillator Design

and is better than a short circuit stub, since the open circuit stub can be trimmed to adjust the circuit. A 50-Ω line is used throughout this circuit (width = 0.071 in for ϵ_r = 2.55 which is a woven PTFE laminate). The value of susceptance is 1.55. This corresponds to a line length of 57.1° (tan^{-1} 1.55). By going through calculations for wavelengths and fractions of wavelengths we find that this length would be 0.284 in. If we use two stubs, one on each side of the main line, we find that the length of each stub is 37.7° (tan^{-1} 0.775). These calculate to be 0.187 in each.

When we travel from point B to point A in Figure 6.7 we travel a distance of 0.379 wavelengths. In order to set a reference point for the transistor a distance of 0.200 in was used between that reference point and the transistor. When this is added ot 0.379λ (0.682 in) we find that the two stubs (or single stub if that is used) are located 0.882 in from the transistor. (Once again this is a 50-Ω line.) The circuit calculated above is shown in Figure 6.8. The circuit is constructed on 0.030 in woven PTFE material and exhibits a power output of 20.45 dB when the transistor draws 30 mA of current.

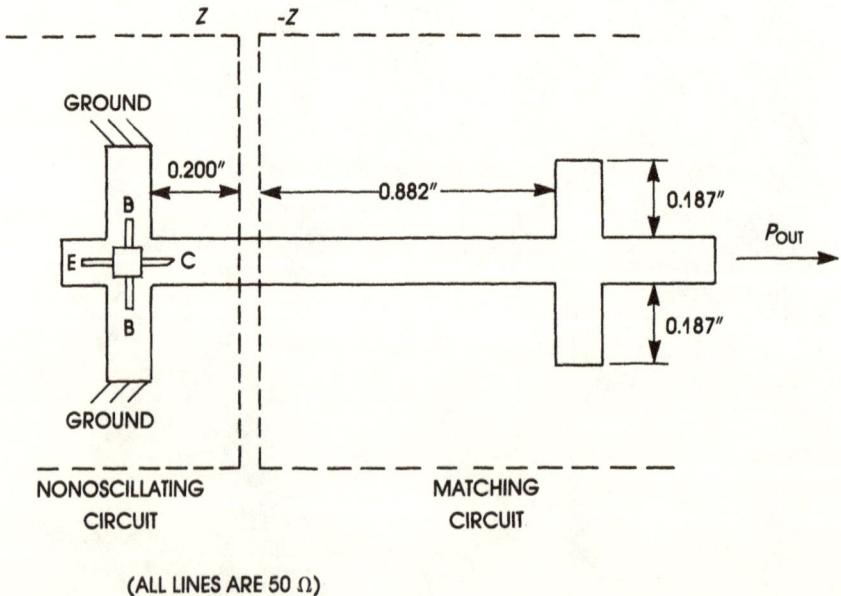

Figure 6.8 Oscillator Circuit

The second type of solid-state oscillator used in microwaves is the Gunn diode oscillator. As we stated in Chapter 4, the Gunn diode is a negative resistance device which exhibits properties that make it ideal for high frequency oscillators. Figure 6.9 shows the current *versus* voltage curve which was presented earlier. You can see how the current increases with voltage to a point (point A on Figure 6.9) where any further increase will cause the current to drop. The best operation is between points A and B where current drop is linear. On either side of these points there are instabilities which will not lend themselves to reliable oscillator operation.

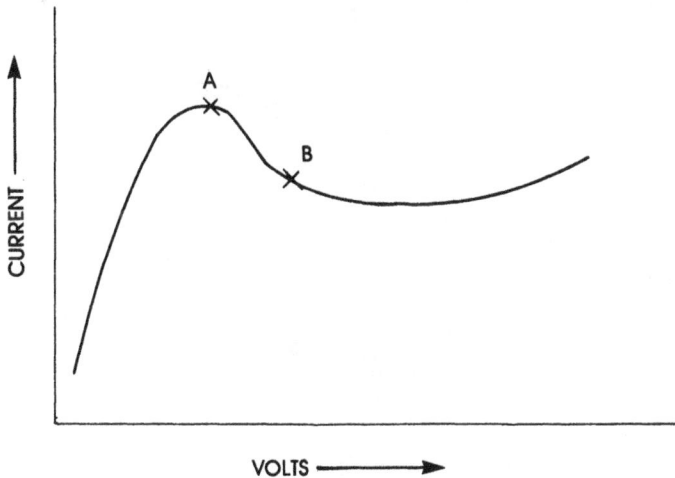

Figure 6.9 Gunn Diode *I-V* Curve

Gunn diode oscillators are generally found in wave-guide circuits because of their frequency of operation. They may be single-diode circuits, double-diode (push-pull) circuits, or may go as high as four diode arrangements when InP diodes are used. Any more than four diodes will cause severe matching problems.

The single Gunn diode waveguide circuit is shown in Figure 6.10(a) and is a half-wave resonant cavity with the diode placed at one end of the cavity. The dc bias for the diode is obtained from an inductive in contact with the diode and RF isolation is obtained by choking techniques. The output power is inductively coupled through an iris with the frequency tunable by a dielectric rod inserted in the cavity.

The double Gunn diode oscillator, or push-pull, waveguide circuit is shown in Figure 6.10(b). With proper coupling it is possible to obtain slightly more than the combined maximum power that each diode is capable of supplying individually. The circuit shown is basically two half-wave single diode waveguide cavities placed end-to-end. The diodes are approximately one guide wavelength apart for proper push-pull operation. With those

conditions, the power can be magnetically coupled to the base in the center
of the cavity. This is ideal since this is the point where the magnetic fields
add in phase. The circuit as shown in Figure 6.10(b) can be tuned with a
dielectric rod slightly off center between the two diodes.

The Gunn oscillator may be a single frequency device, may be a
mechanically tuned oscillator, or it may be electrically tuned by incorporating
a varactor diode (or series of varactor diodes) which will change the circuit
frequency.

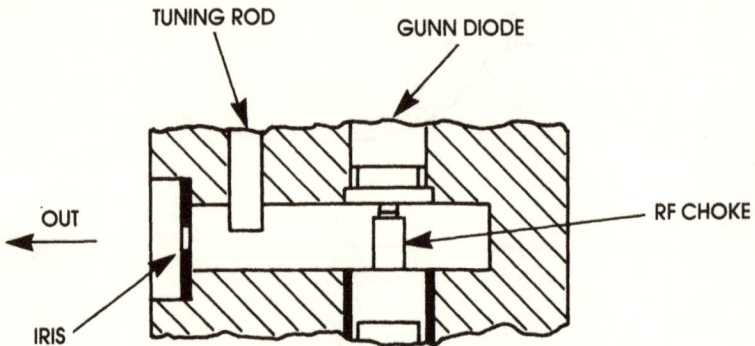

(a) SINGLE DIODE OSCILLATOR

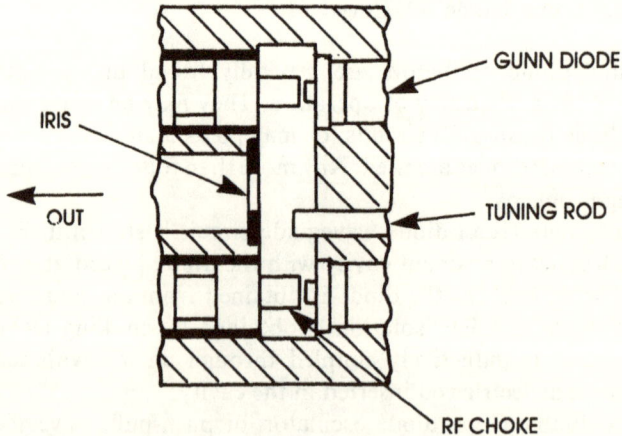

(b) DOUBLE DIODE OSCILLATOR

Figure 6.10 Gunn Diode Oscillators

6.3 ATTENUATORS AND SWITCHES

Most attenuators and switches used in microwaves will incorporate PIN diodes in a variety of configurations. This is a logical choice, since we have already referred to the PIN diode as a microwave variable resistor. We combine the components of attenuators and switches because their operation is very similar. Conduction, and thus attenuation, is controlled in the PIN diode by varying the bias applied to it. If we apply a steadily increasing voltage to a PIN diode circuit, we will have a continuously variable attenuator; a step function with time will result in a step attenuator; and if we use a pulse, or step function, we will obtain a switch. These bias conditions are shown in Figure 6.11. So it can be seen that simply by changing bias conditions or by changing the time of the bias condition we can go from an attenuator to a switch or *vice versa*.

In Chapter 4 we presented a basic quadrature hybrid constant impedance device which could be used as either an attenuator or switch. Figure 6.12 shows this configuration which should be familiar to you. You will recall that in the attenuator mode the input is at port 1, the output at port 4, and ports 2 and 3 are terminated in 50Ω. In the switch mode, the input is still at port 1 with the output at port 4 when the diodes are turned on. When the diodes are off the output is at port 2. Port 4 is now the isolated port.

There are other types of constant impedance components used for attenuators and switches. We will look at three types which should look somewhat familiar to you. They are the π, the T, and the bridged-T configuration. All of these types are used as fixed attenuators when ordinary carbon resistors are placed in the circuits. By replacing the resistors with variable resistors (PIN diodes) we can produce either variable attenuators or very fast switches.

Figure 6.13 shows the π configuration for an attenuator-switch arrangement. It can be seen that this configuration has the familiar series element surrounded by shunt elements. The only difference is that these elements are diodes that can have their resistance changed by varying the bias on them. The value of resistance needed for this configuration is given as follows:

$$R_1 = Z_0 \left[\frac{K+1}{K-1} \right]$$

$$R_3 = \frac{Z_0}{2} \left[K - \frac{1}{K} \right]$$

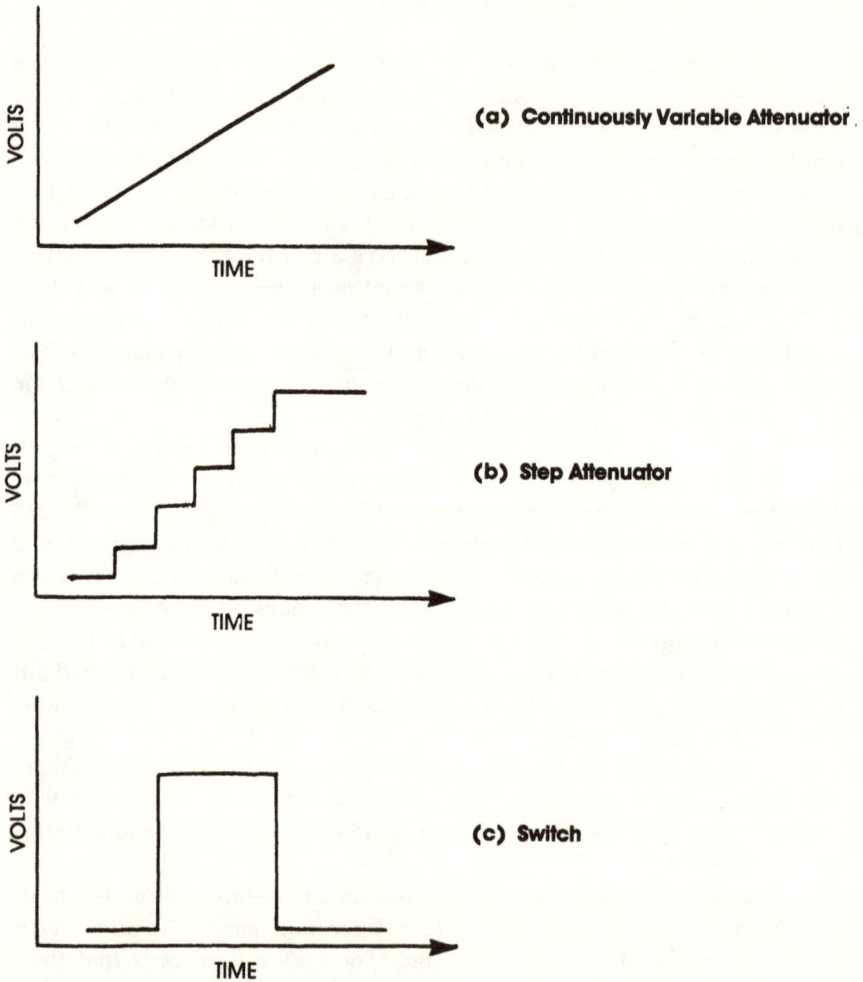

(a) Continuously Variable Attenuator

(b) Step Attenuator

(c) Switch

Figure 6.11 Bias Conditions for a PIN Diode

where

Z = characteristic impedance
K = input to output voltage ratio

It can be seen that by determining what input-output ratio you need, you can adjust the resistors as appropriate to achieve the desired value, or switching characteristics. Two conditions and their resistance values are shown in Table 6.2.

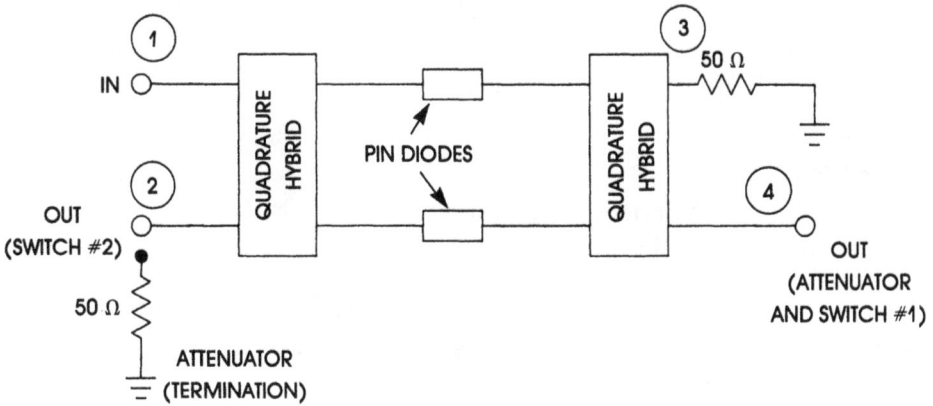

Figure 6.12 Quadrature Hybrid Attenuator and Switch

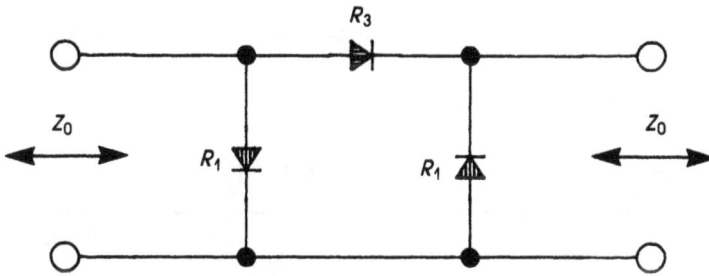

Figure 6.13 The π Configuration

TABLE 6.2

Condition	$R_1 \ (\Omega)$	$R_3 \ (\Omega)$	Theoretical Attenuation (dB)
Isolation	50	10 K	52
Transmission	10 K	1	0

If we were to take a value between isolation and transmission, 10 dB for example, the values of R_1 and R_3 would be 96Ω and 72Ω, respectively.

Figure 6.14 shows the second type of attenuator-switch, the T (or TEE) configuration. Once again, the arangement should be familiar with diodes replacing resistors. Values of R_1 and R_3 are calculated as follows:

$$R_1 = Z_0 \left[\frac{K-1}{K+1} \right]$$

$$R_3 = \frac{2Z_0}{K - \frac{1}{K}}$$

where

Z = characteristic impedance
K = input to output voltage ratio

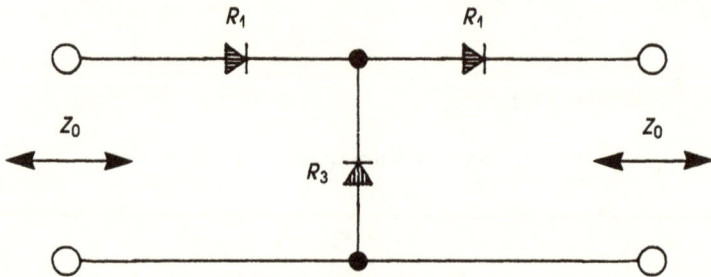

Figure 6.14 The T Configuration

TABLE 6.3

Condition	$R_1 \Omega$	$R_3 \Omega$	Theoretical Attenuation (dB)
Isolation	50	1	40
Transmission	1	10 K	0

If we were to want an attenuation value of 20 dB for this configuration, R_1 would be calculated as 61Ω and R_3 would be 10Ω.

The final attenuator-switch combination we will look at here is shown in Figure 6.15. This is the bridged-T configuration. This arrangement uses only two diodes with two fixed-value resistors. The resistance values used in the bridged-T are calculated as follows:

$$R_1 = R_2 = Z_0$$
$$R_3 = \frac{Z_0}{K-1}$$
$$R_4 = Z_0(K-1)$$

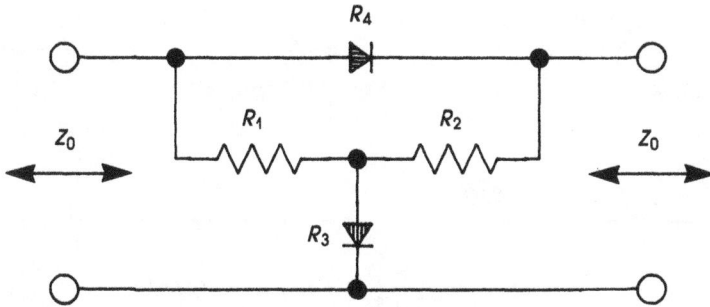

Figure 6.15 The Bridged-T Configuration

where

Z = characteristic impedance
K = input to output voltage ratio

As with the other configurations, representative values of R_1, R_2, R_3, and R_4 are shown in Table 6.4.

TABLE 6.4

Condition	$R_1\Omega$	$R_2\Omega$	$R_3\Omega$	$R_4\Omega$	*Theoretical* Attenuation (dB)
Isolation	50	50	1	10 K	40
Transmission	50	50	10 K	1	0

This configuration also allows values between the attenuation numbers shown. It only requires that you adjust R_3 and R_4 appropriately.

The circuits presented have been ideal representations of π, T, and bridged-T circuits. Figures 6.16 and 6.17 show actual circuits that can be used as variable or step attenuators, or as switches to switch microwave energy on and off. Figure 6.16 is a π attenuator and Figure 6.17 is a bridged-T arrangement. Both have these bias circuits with them to show the complexity and completeness of the designs.

6.4 DETECTORS AND MIXERS

The solid-state components, detectors and mixers, are grouped together because they are the primary users of Schottky diodes. Many of the properties

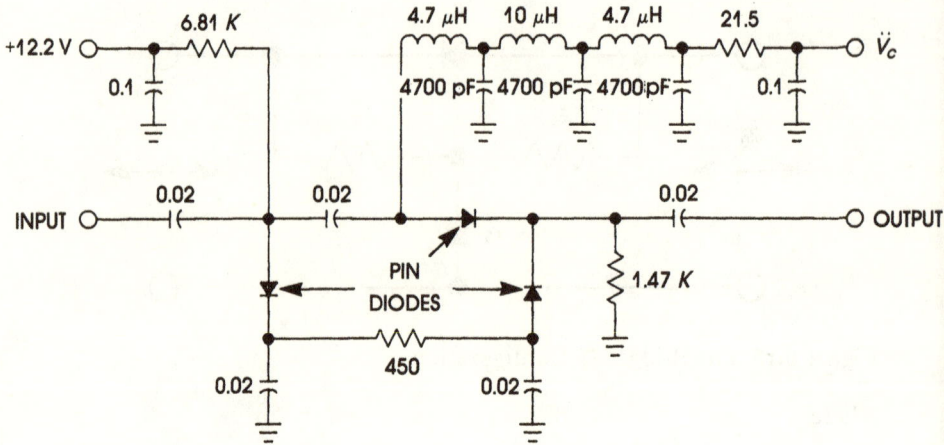

Figure 6.16 The π Attenuator

and applications or detectors and mixers were covered in Chapter 4. Some of these will be repeated here so that our discussions can be complete.

As we stated previously in Chapter 4, the microwave detector is made up primarily of a matching circuit, the diode used, and an output bypass capacitor or filter arrangement to eliminate any RF from the output. An additional element to consider is a dc return for the diodes. We will now look briefly at each one of the elements which make up a microwave detector.

The first element we will look at is the input matching network which will match the diode impedance to the characteristic impedance being used (usually 50Ω). To begin our design we will assume that the diode has an impedance, Z_0, of 20 -j 60Ω (0.4 -j 1.2). (The frequency of operation will be 5.5 GHz.) This is point I on the Smith chart of Figure 6.18. We then move around the VSWR circle until the real axis is reached (the horizontal line through the center of the chart). This distance moved is 0.5λ - 0.356λ or 0.144λ. At this point we will put a quarter-wave transformer. The impedance of the 0.144λ line is determined from the real axis value. This is 0.17 (50Ω or 8.5 ohms. It is this value which is to be matched to 50 ohms. If we use a microstrip construction with ϵ_r = 10.2 it will be 0.290 in wide. The effective dielectric constant will be 8.712. The length of the 8.5 transmission line will be calculated as follows:

$$\lambda = \frac{3 \times 10^{10}}{5.5 \times 10^9 \ \sqrt{8.712}}$$

$$\lambda = 1.848 \text{ cm} = 0.727 \text{ in.}$$

$$0.144 \ \lambda = 0.105 \text{ in}$$

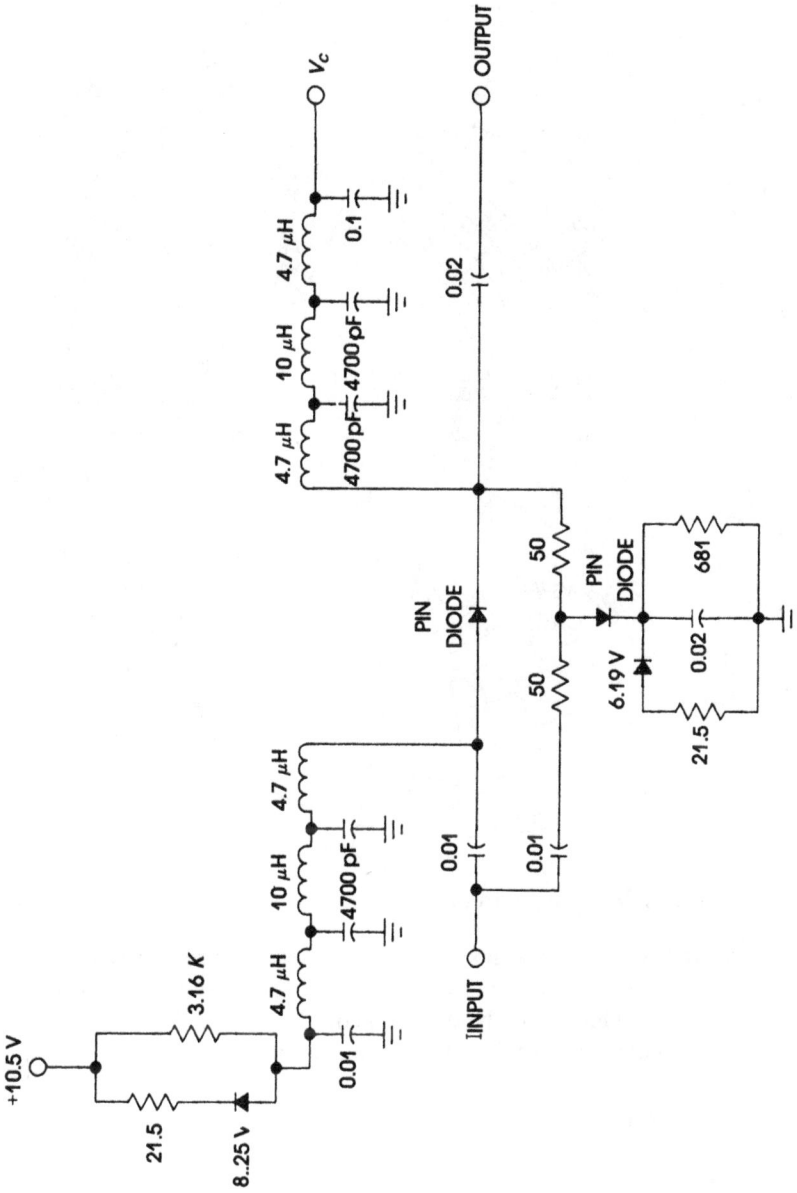

Figure 6.17 The Bridged-T Attenuator

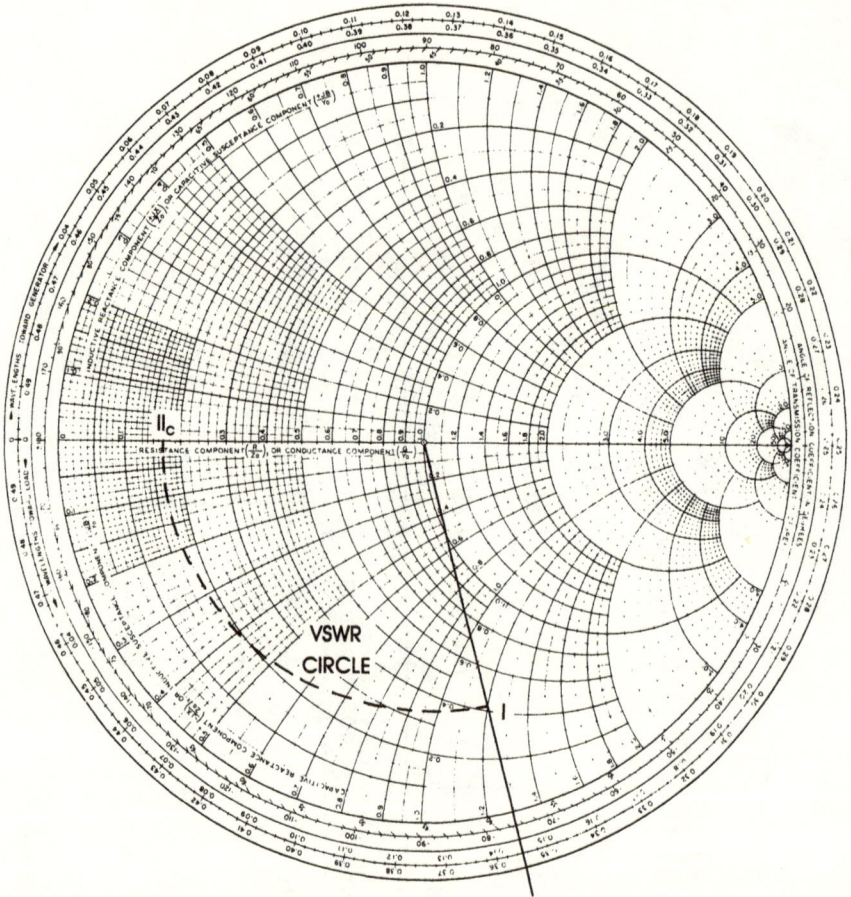

Figure 6.18 Detector Matching Circuit

The quarter-wave transformer is next to be calculated. As previously mentioned, the values to be matched are 8.5 Ω and 50 Ω. There is, therefore, an intermediate value which must be determined, Z_{02} (the 8.5 Ω is Z_{01}, and 50 Ω is Z_0). The calculations of this value are shown below.

$$Z_{02} = \sqrt{Z_0 Z_{01}}$$

$$Z_{02} = \sqrt{50 \, (8.5)}$$

$$Z_{02} = 20.6 \, \Omega$$

The quarter-wave transformer arrangement is shown in Figure 6.19(a).
A quarter wavelength at 5.5 GHz for a 20.6 Ω line is

$$\lambda/4 \;=\; \frac{C}{4f\sqrt{\epsilon_{\text{eff}}}} \;=\; \frac{3\times10^{10}}{4(5.5+10^9)\;\sqrt{7.913}}$$

$$\lambda/4 \;=\; 0.485 \text{ cm} \;=\; 0.191 \text{ in.}$$

Thus, the complete input matching network will look like Figure 6.19(b).

The diode to be used for such a detector should be a Schottky diode which exhibits the impedances we chose (20 –j60 Ω) and will need to be in a package which will exhibit a small value of package capacitance, C_P. Thus, a glass package will *not* be acceptable for this type of application.

The dc return to be used with our detector will be a line printed on the circuit board. This line should be around 100 Ω (0.003 in on our material) and will be one-quarter wavelength long at our frequency of operation (5.5 GHz). This length turns out to be 0.216 in. This line is attached to the diode at one end and to ground at the other.

A bypass filter, or filter arrangement, is designed to be a low-pass network which will only pass the detected voltage and eliminate any RF energy that may get through the detector. A bypass capacitor in the order of 100 pF will short out any RF which get through the 5.5 GHz detector. When a low-pass filter is to be used, the designer can build a one or two pole series inductance-shunt capacitance low-pass filter. This can be fabricated with relative ease on the 10.2 dielectric material we have chosen.

The mixer generally used in microwave applications is the double-balanced mixer. A basic block diagram is shown in Figure 6.20. The input coupling network usually is made up of a quadrature hybrid coupler. This gives the input circuit excellent isolation between the RF and LO signals and provides output to the diode circuits which are equal in amplitude and 90° out of phase.

The input filters, as they are classified in Figure 6.20, may many times be the matching networks which match the input coupling network to the diode circuits. Other times they are actual high-pass filters which will pass the RF and LO frequencies, but not the IF.

The diode circuits may be a single diode, a pair of diodes, a quad of diodes, or other arrangements which are dictated by a particular application or set of specifications. Whatever arrangement is used, the impedance of that arrangement must be known in order to match it to the input at the required frequencies.

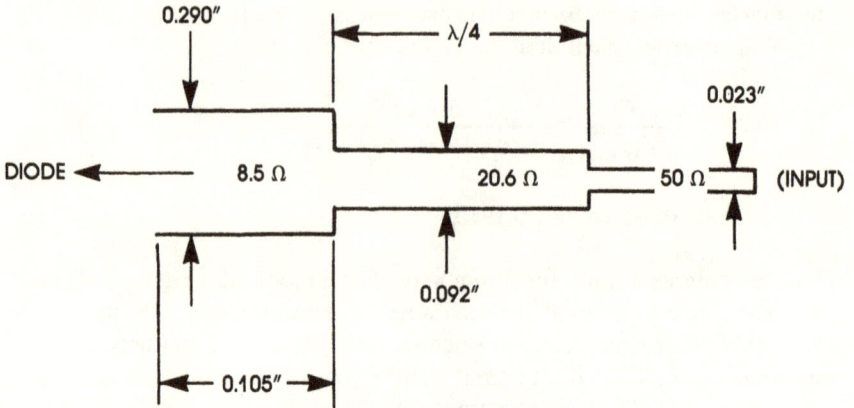

(a) Quarter-Wave Transformer

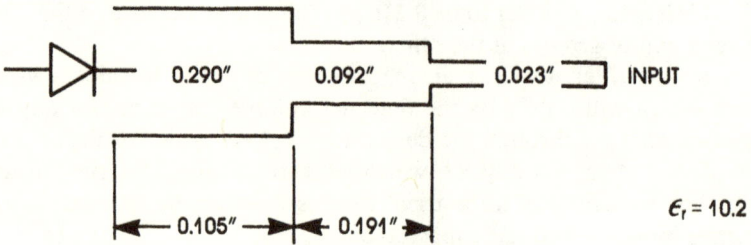

(b) Complete Matching Circuit

Figure 6.19 Transformer and Complete Matching Circuit

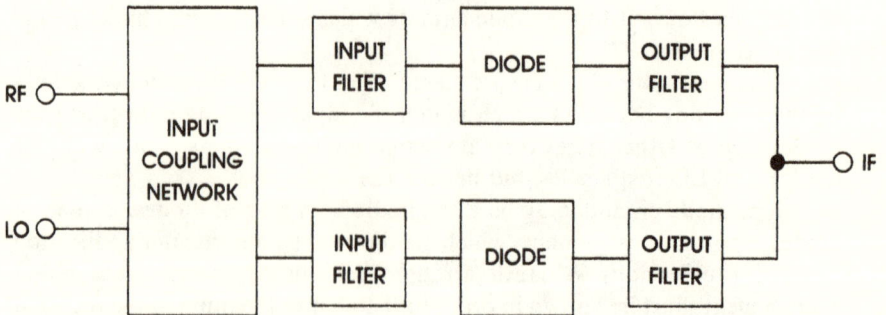

Figure 6.20 Balanced Mixer

The output filters are to be designed to pass the IF and eliminate the RF and LO signals. These filters are usually low-pass filters, since the IF is usually the difference frequency of the RF and LO. The final output is the combination of the two output filters which are connected together. This is necessary since each branch of the balanced mixer produces one-half of the desired IF output.

6.5 PHASE SHIFTERS

A solid-state phase shifter, like any phase shifter, must change the propagation phase of a microwave signal. This task is performed by varying the bias on a diode to switch elements in and out of the circuit. Since time increases in a positive sense, any delay or phase shift must be a negative phase as shown in Figure 6.21. This confirms the fact that the phase of a wave which is delayed with respect to a reference, or input, wave, will have a negative phase. In Figure 6.21 this phase delay (shift) is $\Delta\phi$.

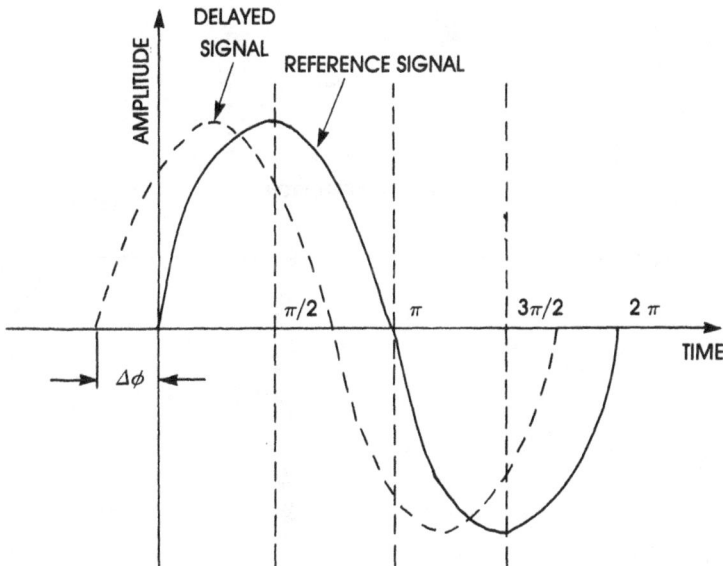

Figure 6.21 Phase Shift

The majority of phase shifters used are *real-time* phase shifters. Two such phase shifters are shown in Figure 6.22. The first, a circulator type, is probably the simplest to design. It uses a circulator, a diode switch, and

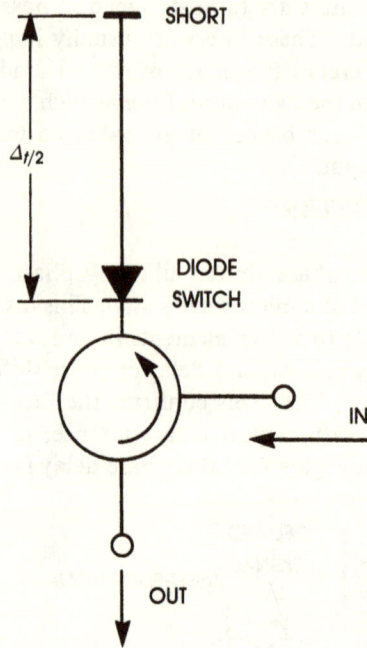

(a) Circulator Type

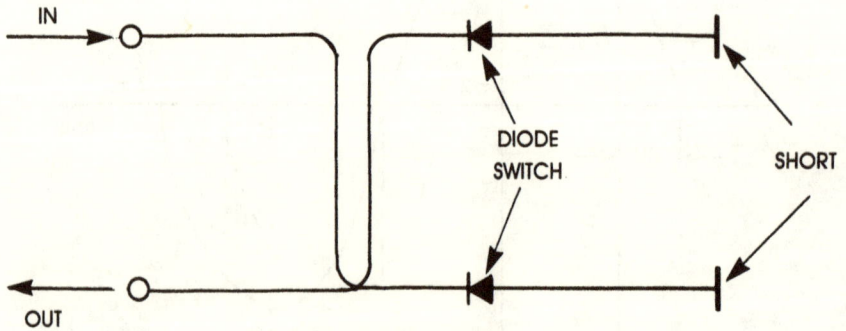

(b) Hybrid Type

Figure 6.22 Real-Time Phase Shifters

a shorted transmission line whose length is equal to $\Delta t/2$. This term is one-half the total time delay needed to provide the required phase shift. This should be understandable since the microwave signal enters the circulator, is sent down the transmission line (if the switch is activated), encounters the short circuit, is reflected back (a distance of twice the line length has now been traveled), and is sent out the circulator output port. If the switch is not activated the energy will simply travel around the circulator with no significant delay other than what the circulator itself may have.

The second circuit in Figure 6.22 replaces the circulator with a quadrature-hybrid coupler. Its operation is similar to that of the circulator circuit. Microwave energy is applied at the input; it divides in the coupler and is seen at the coupler output ports (one half at each port); the energy travels down the transmission lines of the switches activated, is reflected back after it encounters the shorts, and is sent out the output port as a delayed signal. No energy is sent back to the input because of the properties of the quadrature hybrid. With the switches turned off, the energy is all reflected by the open-circuited switches to the output.

A type of dispersive phase shifter is shown in Figure 6.23. This method introduces energy dispersion by periodically shunt-loading a uniform transmission line. It can be seen in Figure 6.23 that this loading is accomplished by using a susceptance jb. The inductance, L, is the total series-equivalent inductance of the diode used and its holder or associated circuitry. This inductance should be measured when the diode is forward biased. The capacitance, C, is obtained when the diode is reverse biased. (C represents the reverse-bias capacitance of only the diode wafer.) The values of L and C can be transformed to any desired values at the junction of the main transmission line by using the diode to terminate a shunting section of transmission line whose length and impedance can be adjusted to produce the desired susceptance. This type of phase shifter is one which finds many applications where frequency-independent phase change is required.

6.6 SUMMARY

Some of the solid-state components have been covered in this chapter. We have chosen some representative examples of amplifiers, oscillators, attenuator-switch combinations, detectors, mixers, and phase shifters. Not every example was covered; obviously, this would probably take more than one volume to accomplish. Only the circuits that were felt to be appropriate were chosen. These examples should give you a start to find that one circuit which is right for your application.

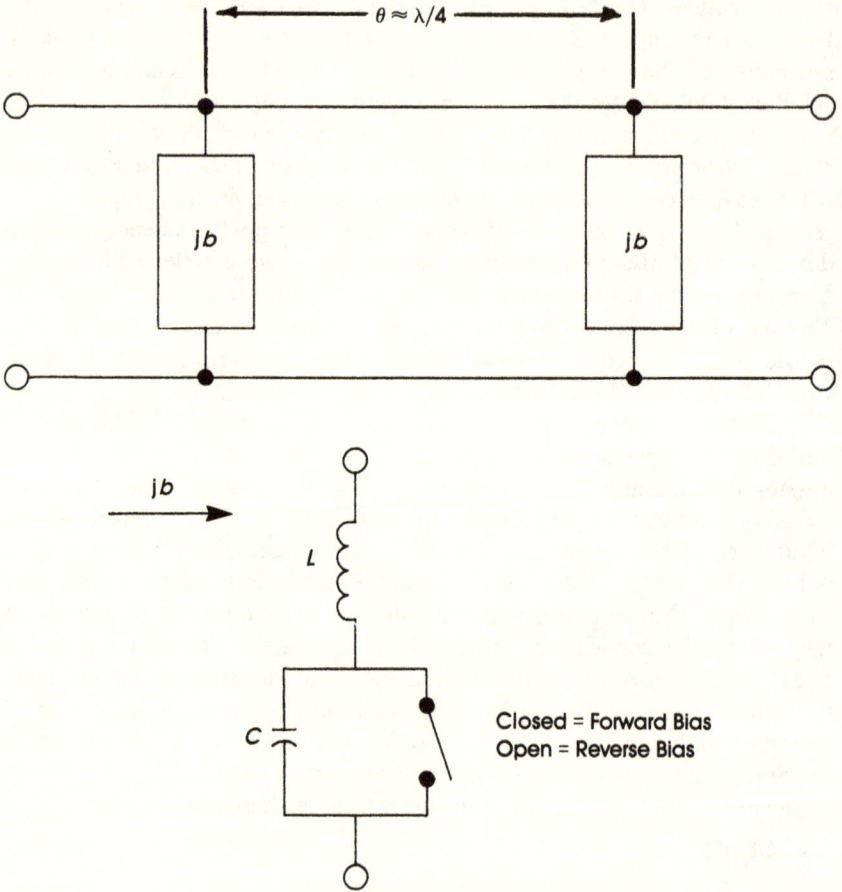

Figure 6.23 Diode-Loaded Transmission-Line Phase Shifter

Glossary

Acronym	Definition
ALC	Automatic Leveling Circuit
CW	Continuous Wave
FET	Field Effect Transistor
FM	Frequency Modulation
GaAs FET	Gallium Arsenide Field Effect Transistor
HEMT	High Electron Mobility Transistor
IF	Intermediate Frequency
IMPATT	Impact Avalanche and Transit Time
IGFET	Insulated-Gate Field Effect Transistor
JFET	Junction Field Effect Transistor
LEC	Liquid-Encapsulated Czochralski
LEK	Liquid-Encapsulated Kyropoulus
LO	Local Oscillator
LPE	Liquid-Phase Epitaxy
LSA	Limited Space-Charge Accumulation
MBE	Molecular-Beam Epitaxy
MLEC	Magnetic Liquid-Encapsulated Czochralski
MESFET	Metal-Semiconductor Field Effect Transistor
MOSFET	Metal-Oxide-Semiconductor Field Effect Transistor
PIN	Positive-Intrinsic-Negative
PTFE	Polytetrafluoroethylene
RF	Radio Frequency
RIMPATT	Read Impact Avalanche and Transit Time
TDA	Tunnel Diode Amplifier
TRAPATT	Trapped Plasma Avalanche Triggered Transit
TSS	Tangential Signal Sensitivity
SPST	Single Pole, Single Throw
VPE	Vapor-Phase Epitaxy
VSWR	Voltage Standing Wave Ratio

Bibliography

CHAPTER 1

Graf, Rudolf F., *Modern Dictionary of Electronics*, Fifth Edition, Howard W. Sams, Indianapolis, 1982.

"1974 Microwave Engineers' Handbook and Buyers Guide," *Microwave Journal*, January 1974.

"The Microwave System Designer's Handbook," *Microwave Systems News*, Vol. 14, No. 7, July 1984.

CHAPTER 2

Chen, Daniel, and D. P. Sui, "Advanced Solid-State Component Design Looks to Gallium Arsenide," *Microwave System Designer's Handbook*, 1984.

Ferry, David K., Ed., *Gallium Arsenide Technology*, Howard W. Sams, Indianapolis, 1985.

Fitchen, Franklin C., *Transistor Circuit Analysis and Design* , D. Van Nostrand, Princeton, NJ, 1966.

Fogell, M., *Modern Microelectronic Circuit Design, I C Applications, Fabrication Technology*, Vols. I and II, Research and Education Association, 1981.

Johnson, V. A., *Men of Physics Karl Lark-Horovitz*, Pergammon Press, Oxford, 1969.

Laverghetta, Thomas S., *Microwave Materials and Fabrication Techniques*, Artech House, Dedham, MA, 1984.

Laverghetta, Thomas S., *Practical Microwaves*, Howard W. Sams, Indianapolis, 1984.

Pearce, C. A., *Silicon Chemistry and Applications*, The Chemical Society, London, 1972.

Riddle, Robert L., and Marlin P. Ristenbatt, *Transistor Physics and Circuits*, Prentice-Hall, Englewood Cliffs, NJ, 1958.

Watson, H. A., *Microwave Semiconductor Devices and Their Circuit Application*, McGraw-Hill, New York, 1969.

CHAPTER 3

Fitchen, Franklin C., *Transistor Circuit Analysis and Design*, D. Van Nostrand, Princeton, NJ, 1966.

Graf, Rudolf F., *Modern Dictionary of Electronics*, Fifth Edition, Howard W. Sams, Indianapolis, 1982.

Laverghetta, Thomas S., *Practical Microwaves*, Howard W. Sams, Indianapolis, 1984.

Kennedy, George, *Electronic Communications Systems*, Second Edition, McGraw-Hill, New York, 1977.

Rose, Robert M., Shephard, Lawrence A., and John Wulff, *The Structure and Properties of Materials*, Vol. IV, Electronic Properties, John Wiley and Sons, New York, 1966.

Watson, H. A., *Microwave Semiconductor Devices and Their Circuit Applications*, McGraw-Hill, New York, 1969.

White, Joseph F., *Microwave Semiconductor Engineering*, Van Nostrand Reinhold, New York, 1982.

GE Transistor Manual, Seventh Edition, The General Electric Co., Syracuse, NY, 1964.

RCA Transistor Manual, RCA Semiconductor and Materials Division, Somerville, NJ, 1962.

CHAPTER 4

Graf, Rudolf F., *Modern Dictionary of Electronics*, Fifth Edition, Howard W. Sams, Indianapolis, 1982.

Laverghetta, Thomas S., *Practical Microwaves*, Howard W. Sams, Indianapolis, 1984.

Watson, H. A., *Microwave Semiconductor Devices and Their Circuit Applications*, McGraw-Hill, New York, 1969.

White, Joseph, *Microwave Semiconductor Engineering*, Van Nostrand Reinhold, New York, 1982.

A Practical Guide to Microwave Semiconductors, Microwave Associates, Burlington, MA, Reprinted from *MSN*, Palo Alto, CA.

Receiving Diode Handbook, Microwave Associates (now M/A-Com), Bulletin 4006, Burlington, MA, 1980.

PIN Diode Designers' Guide, M/A-Com Silicon Products, Catalog 4013, Burlington, MA, 1983.

Microwave Semiconductor, Alpha Industries, Woburn, MA.

CHAPTER 5

Bering, John J., TRW, "HEMT Technology Gains on MM Waves," *Microwaves and RF*, November 1985.

Cook, Harry F., "Introduction to Microwave FET's," Varian Electron Device Group, Santa Clara, CA, 1982.

Derewonko, Henri, and Daniel Delagebeudeuf, Thomson-CSF, "Thomson Continues Work on 18 to 40 GHz TEGFET's," *Microwave and RF*, November 1985.

Graf, Rudolf F., *Modern Dictionary of Electronics*, Fifth Edition, Howard W. Sams, Indianapolis, 1982.

Hamilton, Robert J., and Northe K. Osbrink, Avantek, Inc., "A GaAs FET Primer: Understanding These Vital Devices," *Microwave Systems News*, October 1982.

Laverghetta, Thomas S., *Practical Microwaves*, Howard W. Sams, Indianapolis, 1984.

Swanson, Alan, Herb, John, Ming and Yung, Gould Research Center, San José, CA, "First Commercial HEMP Challenges GaAs FET's," *Microwave and RF*, November 1985.

Transistor Designer's Guide, Microwave Associates (now M/A-Com), Bulletin #5210, 1978.

Applications of Microwave GaAs FET's, Application Manual AN82901-1, California Eastern Laboratories, Santa Clara, CA.

CHAPTER 6

Myers, Dr. Fred A., "Efficient InP Gunn Diodes Shrink Power Requirements," *Microwaves*, May 1980.

Ruttan, Thomas G., and Robert E. Brown, Varian, Palo Alto, CA, "High Frequency Gunn Oscillators," Technical Paper Reprinted #SSW-104.

Saad, Theodore, "The Microwave Mixer, A Technical Discussion," Sage Laboratories, Natick, MA, 1966.

Tenenholtz, Robert, "The Video Detector, A Technical Discussion," Sage Laboratories, Natick, MA, 1968.

Watson, H. A., *Microwave Semiconductor Devices and Their Circuit Application*, McGraw-Hill, New York, 1969.

"12 GHz Amplifier Designs Using the HFET2201," Hewlett Packard Application Note #973, Hewlett Packard, Palo Alto, CA, 1980.

"A 4.3 GHz Oscillator Using the HXTR-4101 Bipolar Transistor," Hewlett Packard Application Note #975, Hewlett Packard, Palo Alto, CA, 1979.

"Applications of PIN Diodes," Hewlett Packard Application Note #922, Hewlett Packard, Palo Alto, CA.

"RF and Microwave Diode Applications Seminars," Hewlett Packard, Palo Alto, CA, 1973.

Index

www.ingramcontent.com/pod-product-compliance
Lightning Source LLC
Chambersburg PA
CBHW021430180326
41458CB00001B/213